BOTANICAL RESEARCH AND PRACTICES

ORIGANUM

TAXONOMY, CULTIVATION AND USES

BOTANICAL RESEARCH AND PRACTICES

Additional books and e-books in this series can be found on Nova's website under the Series tab.

BOTANICAL RESEARCH AND PRACTICES

ORIGANUM

TAXONOMY, CULTIVATION AND USES

ROGER INGRAM
EDITOR

NOTICE TO THE READER

Library of Congress Cataloging-in-Publication Data

ISBN: 978-1-53619-236-0

Published by Nova Science Publishers, Inc. † New York

Contents

PREFACE

Chapter 1 focuses on taxonomic, cultivation and the pharmacological and other uses of *Origanum* species.

In Chapter 2, *Origanum*'s role in synthesising the metal nanoparticles such as titanium dioxide,palladium, silver, gold, palladium nanoparticles supported on magnetic graphene oxide has been discussed in detail. Hence, researchers are using *Origanum* as a precursor in plant-mediated synthesis.

In the last chapter, the authors discuss *Origanum*, the Turkish spice, which has a history dated back to centuries. This herb has found wide applications due to the presence of primary and secondary metabolites.

Chapter 1 - This chapter is focused on taxonomic, cultivation and the pharmacological and other uses of *Origanum* species. *Oregano* (*Origanum* sp.) is an important medicinal and aromatic plant and constitutes one of the most cultivated aromatic plants worldwide. *Origanum* is a genus of herbaceous perennials and sub shrubs in the family *Lamiaceae*, native to Mediterranean region of Europe, North Africa and temperate Asia, where they are found in open or mountainous habitats. The genus *Origanum* has been divided into 10 sections. *Oregano* plant is 20–80 cm in height with alternate leaves of 1–4 cm. The flowers are purple, long, produced in erect spikes. Many subspecies and strains of *Oregano* have been developed by humans over centuries for their unique flavors or other characteristics. Tastes range from spicy or astringent to more complicated and sweeter.

Oregano sold in garden stores as *O. vulgare* may have a bland taste and larger, less dense leaves, and is not considered the best for culinary uses, with a tasteless remarkable and pungent. Cultivation of *Oregano* under foil provides a significantly higher yield of the herb. The raw material from this variant contains more biologically active compounds. *Origanum* species have recently been of great interest in academia, food industry, pharmaceutical industry and agriculture as potential natural additives to replace synthetic products. *Origanum* has been used in both modern and folk medicine for treatment of many ailments. The extract of *Origanum* species have wide range of biologically active compounds which have demonstrated various pharmacological activities, including antimicrobial, antioxidant, antimalerial, stimulative, expectorant, carminative, choleretic, antispasmodic, anticancer etc. The present chapter summarizes on the origin, taxonomy, cultivation methods to improve its oil yields, biological activity and utilization of various species of this plant.

Chapter 2 - Nanotechnology is gaining pace in the field of Science with each passing day due to various nanomaterials such as nanoparticles, nanotubes, nanorods, and hydrogels. These materials showcase versatile properties compared to their bulk counterparts, such as optical, anti-microbial, and magnetic. Different synthesis routes have been investigated to synthesise nanoparticles such as sol-gel (most common), spinning, pyrolysis, chemical vapour deposition, green synthesis, lithography, milling, sputtering, laser ablation, and thermal decomposition. While in all these synthesis routes, green synthesis is the eco-friendly, non-toxic, cost-effective, safe, and feasible route. Plant-mediated green synthesis is the simplest and most convenient method because of the use of metal salts and part of the plant as the precursors in the reaction. The extract prepared from *Origanum* species is used as a bio reductant because of phenolic compounds, act as a reducing agent in the synthesis process. The phytochemicals are also responsible for the stabilisation of the nanoparticles. In this chapter, *Origanum*'s role in synthesising the metal nanoparticles such as titanium dioxide, palladium, silver, gold, palladium nanoparticles supported on magnetic graphene oxide has been discussed in detail. Hence, researchers are using *Origanum* as a precursor in plant-mediated synthesis.

Chapter 3 - Wounds are a global health concern for the world's major population, and it is influenced by many factors such as the local environment of wounds, untreated acute wounds, ageing, diabetes, and arterial disease, which may delay the healing process of wounds. Around 1% of the population suffers from wounds due to the high economy, degradation of living standard, low literacy standard, lack of exposure to health facilities, and complex and long treatments. Different wound healing therapies are being investigated and developed by the researchers to reduce the healing time of wounds and the complexity of wounds. Ancient traditional wound healing therapies derived from herbs are also gaining attention due to their cost-effectiveness and non-toxicity. In this chapter, the authors will discuss *Origanum*, the Turkish spice, which has a history dated back to centuries. This herb has found wide applications due to the presence of primary and secondary metabolites. *Origanum* is used in different forms such as oil, ointment or extract to reduce the complexities in the wound healing process. The essential oil obtained from different species of *Origanum* contains thymol, carvacrol, p-cymene, thymoquinone, ɣ-terpinene and other compounds which are responsible for the healing properties. Study supporting the use of oregano ointment over its essential oil also has been discussed briefly. Researchers have also evaluated the administration of extracts on animal models for its efficacy in wound healing. Due to the goal of sustainable development, traditional therapies will flourish in the coming future.

In: *Origanum*
Editor: Roger Ingram
ISBN: 978-1-53619-236-0

Chapter 1

ORIGANUM: A BRIEF REVIEW ON TAXONOMY, CULTIVATION AND USES

Prasoon K. Joshi[1,*]
and Shishir Tandon[2,†]
[1]Department of Chemistry, M.B. Govt P.G. College, Haldwani (Nainital) Uttarakhand, India
[2]Department of Chemistry (Agricultural Chemicals Division), CBS&H, GBPUA&T, Pantnagar (US Nagar), Uttarakhand, India

ABSTRACT

This chapter is focused on taxonomic, cultivation and the pharmacological and other uses of O*riganum* species. *Oregano* (*Origanum* sp.) is an important medicinal and aromatic plant and constitutes one of the most cultivated aromatic plants worldwide. *Origanum* is a genus of herbaceous perennials and sub shrubs in the family *Lamiaceae*, native to Mediterranean region of Europe, North

* Corresponding Author's E mail: prasoonjoshi2012@gmail.com.
† Corresponding Author's E mail: shishir_tandon2000@yahoo.co.in.

Africa and temperate Asia, where they are found in open or mountainous habitats. The genus *Origanum* has been divided into 10 sections. *Oregano* plant is 20–80 cm in height with alternate leaves of 1–4 cm. The flowers are purple, long, produced in erect spikes. Many subspecies and strains of *Oregano* have been developed by humans over centuries for their unique flavors or other characteristics. Tastes range from spicy or astringent to more complicated and sweeter. *Oregano* sold in garden stores as *O. vulgare* may have a bland taste and larger, less dense leaves, and is not considered the best for culinary uses, with a tasteless remarkable and pungent. Cultivation of *Oregano* under foil provides a significantly higher yield of the herb. The raw material from this variant contains more biologically active compounds. *Origanum* species have recently been of great interest in academia, food industry, pharmaceutical industry and agriculture as potential natural additives to replace synthetic products. *Origanum* has been used in both modern and folk medicine for treatment of many ailments. The extract of *Origanum* species have wide range of biologically active compounds which have demonstrated various pharmacological activities, including antimicrobial, antioxidant, antimalerial, stimulative, expectorant, carminative, choleretic, antispasmodic, anticancer etc. The present chapter summarizes on the origin, taxonomy, cultivation methods to improve its oil yields, biological activity and utilization of various species of this plant.

Keywords: *Origanum*, review, taxonomy, cultivation, uses

INTRODUCTION

Plants have been used throughout, since ancient time, in cooking, medicines, cosmetics, and insecticides with their volatile oil fraction. The International Organization for Standardization (ISO) officially recognizes approximately 70 herbs and spices, while the number used around the world is probably in excess of 350-400 species (Nakatani, 1994). Many natural compounds obtained from the extracts of plants have biological activities and among these (Tandon et al., 2009; Tandon et al., 2001), volatile oils from aromatic and medicinal plants are particularly interesting and have many interesting phamacological, biocidal and antimicrobial activities as described by many researchers (Deans and Ritchie, 1987; Graven, 1992; Baratta, 1998; Tandon et al., 2005; Joshi et al., 2011; Tandon and Mittal, 2018). In aromatic herbs, the essential oils are largely located within the glandular structures that develop on the surface of leaves as well as other organs of the plants. The peltate hairs contain most of the oil (Bosabilidis and Tsekos, 1984) and are so called 'the glands.' Each gland originates from a single protodermal cell that undergoes division and derives two unequally sized cells. The lower cell corresponds to the foot cell while the upper cell further divides to yield the stalk cell and the mother head cell. The foot and stalk cell remain unicellular throughout the subsequent development of the gland while the mother cell of the head further divides to give rise to eight or 12 head cells by the end of the development.

Origanum (Labiatae) species are widely distributed in world in which *O. majorana* L., *O. onites* L., *O. minutiflorum* L., *O. syriacum* L. and *O. vulgare* L. ssp. *hirtum* (Link) Ietswaart are the main species of oregano. The plants are distributed and cultivated mainly in the Mediterranean region and also in many areas of the mild, temperate climates of Europe, Asia, North Africa and America (Ietswaart, 1980; Goliaris et al., 2002). These species have been used since ancient times in Anatolia as condiments and herbal teas. The production of aromatic crop Oregano is due to the different applications of this crop, such as ornamental cropping or food production, and also as a consequence of the antimicrobial, antifungal, insecticidal and antioxidant effects of its essential oils. Interest among consumers in products derived from aromatic/ medicinal plants is expanding. The growing demand has given rise to an increase in the number of specialized industries that utilize this plant. Harvesting of the wild species of Oregano is not usually carried out using the most adequate methods, as frequently the plants are pulled out of the ground by hand.

The genus *Origanum* contains two important aromatic plants with different sensorial qualities, marjoram (*O. majorana* L.) and Oregano (*Origanum* sp.), but also species from other genera with similar sensorial qualities are called 'Oregano' (Lawrence, 1984). The bicyclic monoterpene *cis*-sabinene hydrate and its acetate derived from the 'sabinyl' pathway are responsible for the quality of marjoram, while the phenolic monoterpene carvacrol, arising from the 'cymyl' pathway (Skoula et al., 1999), is the character compound of oregano (Figure 1).

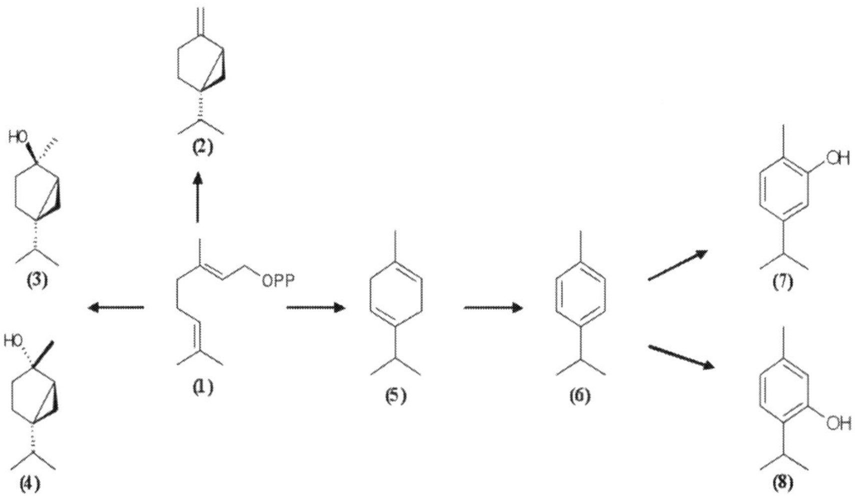

Figure 1. Biosynthesis of 'sabinyl' and 'cymyl'-type monoterpenes in *Origanum*. geranylpyrophosphate (1); 'sabinyl'- types: sabinene (2), *cis*-sabinene hydrate (3), *trans*-sabinene hydrate (4); 'cymyl'-types: gamma-terpinene (5), p-cymene (6), carvacrol (7), thymol (8).

The monoterpene phenols carvacrol and thymol are well known for their high antimicrobial (Sivropoulou et al., 1996; Dorman and Deans, 2004; Elgayyar et al. 2001; Paster et al., 1995) and antioxidant activities (Lagouri et al., 1993; Ai Bandak and Oreopoulou, 2007).

The members of genus *Origanum*, are erect or ascending, perennial and highly aromatic. Based on morphological criteria, Ietswaart (1980) reconized 10 sections, consisting of 42 species or 49 taxa (species, subspecies and varieties) within genus *Origanum*. Before 1980, *O. vulgare* L. indifferently referred the subspecies that Ietswaart later identified as *O. vulgare* L. ssp. gracile (C. Koch) Ietswaart, *O. vulgare* L. ssp. *glandulosum* (Desfon-taines) Ietswaart, *O. vulgare* L. ssp. *hirtum* (Link) Ietswaart, *O. vulgare* L. ssp. *vulgare* L., *O. vulgare* L. ssp. *virens* (Hoffmannsegg et Link) Ietswaart and *O. vulgare* L. ssp. viride (Boiss.) Hayek.

In the Indian subcontinent, *O. vulgare* L. is represented by the typical subspecies, *vulgare*, which is locally known as 'Himalayan Marjoram' and is widely distributed in the subtemperate to temperate regions of Himalaya (Hedge, 1990; Mukerjee, 1940). In India, it is found in Jammu and Kashmir, Himanchal Pardesh, Uttar Pradesh, Uttarakhand, and Sikkim. In northern Himalayan region: *O. Vulgare* found in seven districts of

Uttarakhand situated at different geographical locations (Singh et al. 2018). Dried *Origanum* leaves and essential oils are used by the flavouring industry in preparing various liqueur formulations, tomato sauces, condiments, in baked goods such as pizzas and salad dressings (Ozel and Kaymaz, 2004). The essential oil of this plant has been proven to have antimicrobial and antioxidant activities. Oregano has long been recognized as a culinary herb and medicinal plant with beneficial effects on the digestive and respiratory systems and antiseptic, antispasmodic, carminative and cholagogue properties (Makinen and Paakkonen, 2002; Baricevic and Bartol, 2002; Duke, 1992). An excellent review article on chemical, biocidal and pharmacological aspects of *Origanum* has been published by authors (Joshi et al., 2015). In present chapter, authors highlighted the taxonomic, cultivation and uses of *Origanum*.

A literature search was conducted from Journals/literature subscribed in Library of GB Pant University of Agriculture and Technology, Pantnagar, US Nagar, Uttarakhand, India, Scopus, Research gate, open access journals and Wikipedia. All pictures of *Origanum* and structures of compounds are taken from Google search under creative common license.

TAXONOMY

The family Labiatae/Lamiaceae consists of about 221 genera and 5,600 species comprised of aromatic, annual or perennial herbs, sub-shrubs, or shrubs. The leaves are opposite, sometimes whorled or even spirally arranged, and simple rarely pinnate. The flowers are bisexual, zygomorphic, usually have the floral formula K (5), C (5), A (4), and G (45) and are arranged in verticillasters. The corolla is bilabiate and the stamens didynamous. With the exception of *Rosmarinus*, the genera have a gynobasic style and the four one-seeded nutlets have only a small surface of contact with one another. Moreover, the herbaceous members of the family have square stems. The leaves and other aerial parts may be pubescent and possess glandular trichomes, which contain the essential oil.

The genus *Origano* is classified as follow:

Kingdom	- Plantae	**Subkingdom**	- Tracheobionta	- Vascular plants
Superdivision	- Spermatophyta (Seed plants)	**Magnoliophyta**	- Flowering plants	
Class	- Magnoliopsida (Dicotyledons)	**Subclass**	- Asteridae	
Order	- Lamiales	**Family**	- Lamiaceae	- Mint family
Genus	- *Origanum* L. – *Origanum*	**Species**	- *Origanum vulgare* L.	- oregano

The genus *Origanum* [family: Labiatae/Lamiaceae] includes, in the east Mediterranean, 22 species in Turkey (Ietswaart, 1982; Davis, 1982), and 4 in Greece (Strid and Tan, 1991). The Flora Europaea reports 13 species, distributed in 3 Sections: the *Origanum* Section, the *Majorana* Section and the *Amaracus* Section (Tutin et al., 1972). Other species of *Origanum* in the east Mediterranean belonging to the Majoram, *O. majorana* var. *tenuijolium, O. dubium* and *O. onites* (Arnold et al., 1993). The distributions of some *Origanum* species in all over the world are listed in the given Table 1.

Table 1. Distribution of some *Origanum* species in various countries (Padulosi, 1997)

S. No	Country	Species found
1	Afganistan	*O. vulgare* L. subsp. *viridulum* (Martrin- Donos) Nyman *O. vulgare* L. subsp. *gracile* (Koch)Ietswaart
2	Algeria	*O. floribundum* Munby *O. vulgare* L. subsp. *glandulosum* (Desfontaines) Ietswaart
3	Albania	*O. vulgare* L. subsp. *hirtum* (Link) Ietswaart
4	China	*O. vulgare* L. subsp. *viridulum* (Martrin-Donos) Nyman *O. vulgare* L. subsp. *vulgare*
5	Crotia	*O. vulgare* L. subsp. *hirtum* (Link)Ietswaart *O. vulgare* L. subsp. *viridulum* (Martrin-Donos) Nyman
6	Cyprus	*O. cordifolium* (Montbret & Aucher ex Benth.) Vogel *O. laevigatum* Boiss. *O. majorana* L. *O. syriacum* L. *O.* var. *bevanii* (Holmes) Ietswaart
7	Egypt	*O. isthmicum* Danin *O. syriacum* L. *O.* var. *sinaicum* (Boissier) Ietswaart
8	France	*O. vulgare* L. subsp. *viridulum* (Martrin- Donos) Nyman
9	Greece	*O. calcaratum* Juss. *O. onites* L. *O. scabrum* Boiss. & Heldr. in P.E. Boissier

S. No	Country	Species found
		O. vulgare L. subsp. *hirtum* (Link) Ietswaart *O. vulgare* L. subsp. *viridulum* (Martrin- Donos) Nyman *Origanum microphyllum* (Benth.) Vogel *Origanum sipyleum* L. *Origanum symes* Carlström *Origanum vetteri* Briq. & Barbey
10	India	*O. vulgare* L. subsp. *viridulum* (Martrin- Donos) Nyman *O. vulgare* L. subsp. *vulgare*
11	Iran	*O. vulgare* L. subsp. *viridulum* (Martrin-Donos) Nyman *O. vulgare* L. subsp. *vulgare* *O. vulgare* L. subsp. *gracile* (Koch) Ietswaart
12	Iraq	*O. acutidens* (Hand. - Mazz.) Ietsw.
13	Isreal	*O. ramonense* Danin *O. syriacum* L. var. *syriacum* *O. dayi* Post
14	Italy	*O. onites* L. *O. vulgare* L. subsp. *viridulum* (Martrin- Donos) Nyman
15	Jordan	*O. jordanicum* Danin & Künne *O. petraeum* Danin *O. punonense* Danin *O. syriacum* L. var. *syriacum* *Origanum syriacum* L.
16	Lebanon	*O. ehrenbergii* Boissier *O. libanoticum* Boiss. *O. syriacum* L. *O.* var. *bevanii* (Holmes) Ietswaart
17	Libya	*O. akhdarense* Ietsw. & Boulos *O. cyrenaicum* Bég. & Vacc. *O. pampaninii* (Brullo & Furnari) Ietsw
18	Madeira	*O. vulgare* L. subsp. *virens* (Hoffmannsegg & Link) Ietswaart
19	Morocco	*O. elongatum* (Bonnet) Emberger et Maire *O. grosii* Pau et Font Quer ex Ietswaart *O. compactum* Bentham *O. vulgare* L. subsp. *virens* (Hoffmannsegg & Link) Ietswaart
20	New Zealand	*O. vulgare* L.
21	Pakistan	*O. vulgare* L. subsp. *viridulum* (Martrin-Donos) Nyman
22	Portugal	*O. vulgare* L. subsp. *virens* (Hoffmannsegg & Link) Ietswaart
23	Palestine	*O. syriacum* L.
24	Spain	*O. compactum* Bentham *O. vulgare* L. subsp. *virens* (Hoffmannsegg & Link) Ietswaart

Table 1. (Continued)

S. No	Country	Species found
25	Syria	*O. bargyli* Mouterde *O. laevigatum* Boiss. *O. syriacum* L. *O. syriacum* L. var. *syriacum* *O.* var. *bevanii* (Holmes) Ietswaart
26	Saudi Arabia	*O. syriacum* L.
27	Turkey	*O. acutidens* (Hand. - Mazz.) Ietsw. *O. amanum* Post *O. bargyli* Mouterde *O. bilgeri* P.H. Davis *O. boissieri* Ietswaart *O. haussknechtii* Boiss. *O. husnucan-baseri* H.Duman, Aytac & A.Duran *O. hypericifolium* O.Schwarz & P.H.Davis *O. laevigatum* Boissier *O. leptocladum* Boiss. *O. majorana* L. *O. minutiflorum* O. Schwarz & P.H. Davis *O. munzurense* Kit Tan & Sorger *O. onites* L. *O. rotundifolium* Boiss *O. saccatum* P.H. Davis *O. sipyleum* L. *O. solymicum* P.H. Davis *O. syriacum* L. *O.* var. *bevanii* (Holmes) Ietswaart *O. vogelii* Greuter & Burdet *O. vulgare* L. subsp. *gracile* (Koch) Ietswaart *O. vulgare* L. subsp. *hirtum* (Link) Ietswaart
28	Tunisia	*O. vulgare* L. subsp. *glandulosum* (Desfontaines) Ietswaart
29	USSR	*O. vulgare* L. subsp. *gracile* (Koch) Ietswaart
30	Venezuela	*O. vulgare* L.

O. vulgare L., is widely used as a spice and is an extremely variable species which extends from Macaronesia across a large part of Europe and the Mediterranean to Asia. *O. vulgare* L. ssp. *uiridulum* (Martin Donos) Nyman, commonly known in Greece as “righani” or “aghriorighani” is a synonym of *O. heracleoticum* L., *O. viride* (Boiss.) Hal Pcsy, *O. viridulum* (Martin Donos), *O. vulgare* ssp. *heracleoticum* (L.) Holmboe and O. *vulgare* ssp. *viride* (Boiss.) Hayek (Greuter, 1986). It is a perennial plant, woody, rhizomic, very aromatic, 60 cm high, with hirsute branches in the upper part. The leaves are 15-22 x 6-15 mm; they are oval and elongated, entire and slightly notched, slightly hairy, glandular-punctate, petiolate. The flowers are united in small oblong 5-20 mm spikes which form a panicle. The calyx has 5 equal teeth which are glabrous or hairy, with yellow glands. The *viridulum* subspecies differs from the *mlgare* ssp. in the following characteristics: it has 2-3 mm green, often slightly violet bracts; its outermost surface is densely glandulous; the corolla is white, in rare cases with touches of rose. It is spread throughout southeastern Europe, Corsica, Sardinia, in regions of the Aegean, as far as China (Ozel and Kaymaz, 2004; Greuter et al. 1986; Davis, 1982; Strid and Tan, 1991; Tutin et al., 1972*),* and is common in every part of Greece, especially at low altitude, occasionally at 1500-1800 m. The shrub grows to approximately 30-130 cm and is much ramified. The older branches are covered with a light brown bark, while the younger ones are tetragonal, rounded, glandular and covered with thick long and short hairs; the sub sessile leaves have a short petiole; the bracts are imbricate, densely pubescent or tomentose; the calyx is sheath-shaped and the corolla is white. It has a pleasant, aromatic fragrance, especially during the flowering period from June to December and even after the flowers is dried. It is found everywhere, often in rocky places, in the hills, in steep highlands with calcarious rocks, sometimes on old limestone walls, from 150-1000 m Commonly known as "Za'Atar", historically as the "Hyssop" of the Bible (Mouterde, 1984), it originates in Lebanon and other countries of the east Mediterranean: Asiatic Turkey, Cyprus, Syria, Israel and Jordan as far as the Sinai (Greuter, 1986). The shrub *O. ltbanoticum* Boiss., a synonym of *Amuractrs leptocladus* (Boiss.) Briq., is endemic of Lebanon, it is very ramified with glabrous branches. It grows in humid, rocky, montane valleys on the calcareous rocks often in partially shady zones, in grassy places and open forests of Lebanon at altitudes between 400 and 1500m. Its flowers grow in the spring and summer. According to the classification of the species *O. vulgare* L., six sub-species are included: *gracile, hirtum, viride, virens, vulgare* and *glandulosum* (Ietswaart, 1980; Lawrence, 1984). *O. vulgare* ssp. *hirtum* is the Greek oregano, one of the most important commercial oreganos (Ietswaart, 1980; Lawrence, 1984; Lawrence, 1989; Kokkini and Vokou, 1989; Sezik et al., 1993; Baser, 1994; Melegari et al. 1995; Lawrence, 1995; Kokkini et al., 1997; Antuono et al., 2000). *O. vulgare* ssp. *vulgare* (wild marjoram) is the most commonly found oregano in Europe (Maarse and Van Os, 1973; Nykänen, 1986; Sezik et al., 1993; Melegari et al. 1995; Chalchat and Pasquier, 1998; Antuono et al., 2000; Pande and Mathela, 2000 Mockute et al., 2001). The name *Origanum* is often improperly used to designate other species, including Spanish *origanum* (thymus *cupitatus)* (Lawrence, 1993) and Mexican *Origanum (Lrppia graveolm)* (Ravid and Putievsky, 1986). The commercial name 'oregano' refers to a number of species that produce essential oils with a characteristic smell due to the high concentration of carvacrol (Padulosi, 1997). Among the most popular oregano spices are the commercially known as Greek oregano [*O. vulgare* L. ssp. *hirtum* (Link) Ietswaart] and Turkish oregano (*O. onites* L.). *O. vulgare* ssp. h*irtum* (section *Origanum*) occurs mainly in the Balkan Peninsula, Turkey and Cyprus, between sea level and 1,500 m (Ietswaart, 1980). *O. onites* [section *Majorana* (Miller) Bentham] has a rather narrow distribution, restricted in Greece and in west and south Turkey; It grows at altitudes between sea level and 1,400 m (Ietswaart, 1982) *O.* x *majoricum* Cambessedes, which is not listed by Ietswaart to be found in France, was harvested in southeastern France. It is 'false marjoram,' a sterile hybrid, reputedly a cross between the common oregano, *O. vulgare* ssp. *virens*, and marjoram, *O. majorana*. It is cold-resistant, with a survival rate of 80% after harsh winters. It grows rapidly and blooms abundantly. *O. majorana* L. is a native plant of Turkey. It is naturally distributed in an area stretching from Antalya to Içel provinces in the southeastern Mediterranean region of Turkey. This species is characterized by high carvacrol content in its essential oil (Baser et al., 1993).

O. vulgare L., commonly known as Himalayan marjoram in India, is an erect perennial aromatic herb with small, pale pink flowers crowded into a branched domed inflorescence. This plant is 20–80 cm high with ovate entire and staked leaves measuring 1–4 cm. It is found in temperate regions from 7000 to 12,000 ft. It is also found in Pakistan, Bhutan and temperate Eurasia at altitudes of the range 1500–3600 m (Polunin and Stainton, 1984), usually in chalky soil (Ietswaart, 1980). Images of some *Origanum* species are shown in Figure 2.

Figure 2. Images of some species of *Origanum*.

CULTIVATION

The research on cultivation of *Origanum* to meet the demand for use in food and phytotherapy, and therefore to establish the quality of its oil and identify harvest techniques that would make sustainable production possible.

Soil Type and Soil pH

Origanum is a perennial growing plant. It can grow in semi-shade or no shade. It prefers dry or moist soil. The Greek oregano grows in almost every soil type, as long as it has excellent drainage. This fact can be assessed as positive, especially when the oregano crop is to be grown on poor and infertile soils. With regard to pH, oregano can be cultivated in slightly acidic, neutral and slightly alkaline soils, i.e., Oregano is developed in a pH limit between 6.0 and 9.0, with a favored range between 6.0 and 8.0 (Peter, 2004). However, for the production of high-quality dried herb and essential oil, the oregano crop should be grown in calcareous soils of semi-mountainous areas, with cool summers, while the soil pH should be between 6.5 and 7.5.

Fertility

The nitrogen requirement for cultivation of *Origanum* is between 80 and 120 kg/ ha and applied in split dose i.e., 50% pre plant and 50% in two or more side dresses during growing season.

Physical Characteristics

Temperature Range

With respect to the temperature range, Greek oregano plant withstand extremely low temperatures ranging -23 to -29^{0}C, and high ones ranging from 40 to 42^{0}C. However, the most suitable areas for growing Greek oregano are those where temperatures range from 4 to 33^{0}C, while the optimum temperature range lies between 18 and 22^{0}C.

Flowering, Fruiting and Harvesting Time

The species is hermaphrodite (has both male and female organs) and is pollinated by bees. Its flowers bloom from June to September, and the seeds ripen from August to September. It is noted for attracting wildlife.

Harvesting in June to August

The exact time of harvesting is determined on the yield quality basis, e.g., essential oil content upon flowering (usually mid-June to late August). Harvesting is normally carried out mechanically. The plants are cut 5-8 cm above ground. If the crop is irrigated, a second harvest may be possible; in this case, irrigation is needed immediately after the first harvest. The post-harvest processing is occasionally manual. Oregano is harvested at full blooming for essential oil production or at the beginning of flowering for herb production. Significant differences in yield and oil content have not been observed during the flowering period. Oil content in leaves is very low in the autumn (October) harvest. Highest yields are obtained in the second year, during the two cuts made in June-July and in October. Dry oregano yields may reach 1.5-3.0 t per ha, whereas the total duration of Greek oregano cultivation may exceed 10- 11 years. A rough estimate of an average crop production indicates a value of 20 t per ha in a 4-year cultivated field. A full crop performance may be attainable after the first year of establishment. The harvest index is about 50-55% in the first harvest owing to the high incidence of stems and ca. 60-70% in the second mowing usually done in October. Finally, if oregano is intended for the market as fresh or frozen product, the plants can be harvested every 5-6 weeks. In this case, the upper plant parts are cut before flowering can be set. The drying method plays major role for producing a high-quality product. During the drying process, drying temperatures are 30-40 °C should not be exceeded as they negatively affect the oil content (Marzi, 1996). The final product consists of a mixture of leaves and flowers, which account for about 40-60% of the aerial part of the dried plants. The relative volume by weight (specific gravity) of the fresh air portion is 80-120 kg per m (Katsiotis and Chatzopoulou, 2010). Direct exposure to sun is rather dangerous, drying should take place in shade by spreading the harvested material in thin coats and applying ventilation of dry air. Another method is drying by heating. It is used on an industrial level where moisture is dropped by discharging hot air. An alternative form of artificial drying that is well used in the food industry for effective drying aromatic and medicinal plants is freeze drying/lyophilization. The product is cooled and then dried by sublimation of ice in a vacuum. The advantage of this method is the stability of the plant material properties, but its substantial cost comprises a serious drawback.

Various agronomic and physiological experiments were done for knowing the proper quality and quantity of the plant. The wild marjoram seeds cultivated in the Experimental Garden of Tartu University were investigated on the influence of cut-off levels on the quality and quantity of the essential oils, the β-caryophyllene was reported as major compound ranged between 20-45%. (Ivask et al. 2005). The wild oregano species can grow at different altitudes from coastal to mountainous areas in undulating to rolling terrains on various soils with different fertility status, able to withstand rather low temperatures, and finally produce high yields of good quality. High dry matter yields are associated with increased NPK dressings (Goliaris, 1997b). According to Dordas et al. (2009) and Dordas, (2012) the nitrogen is an important element for increasing dry matter yields without affecting the essential oil content significantly. Oregano plants remove from the field 25-86 kg N per ha.

The oil compositions from two different *Origanum* species of two different chemotypes (a "mixed" type with both pathways (sabinyl/p-cymyl) and a pure p-cymyl chemotype raised at different temperatures. Cuttings were prepared from two single plants of *O. vulgare* x *O. majorana* cv. 'Kaliteri' (Richters, Goodwood, Canada) and *O. syriacum* ssp. *syriacum* and potted in 17 cm x 17 cm pots. The plants were grown in triplicate in growth chambers at three different temperature levels (18°C, 22°C and 26°C) kept constant at day and night under long-day conditions with 16 h day length. The major 'sabinyl' compound *cis*-sabinene hydrate was not influenced by temperature, while temperature significantly influenced thymol and carvacrol and other essential oil compounds. Although thymol and carvacrol are closely related monoterpene phenols, they reacted to varying levels of temperature in an opposite way. Thymol increased with decreasing temperatures while carvacrol increased with increasing temperatures. (Novak et al. 2010). Micropropagated plants of *O.*

bastetanum were obtained from seeds of wild specimens collected in Benamaurel (Granada) during the ripening period. Germination, propagation, growth and acclimatization were carried out according to the method described by Socorro et al. (1988). Subsequently, the micropropagated specimens were transplanted into four different plots in Lanjarón (Granada, Spain), and cultivated for three years. These were collected during the flowering period. The samples corresponded to three different cut-off levels: 25%, 50% and 75% of the aerial parts. Oils were obtained by hydro distillation using a Clevenger-type device from the three samples (cut-off level = 25%, 50% and 75% of aerial parts) of *O. bastetanum* cultivated specimens. The oil yields were very similar for the for the three cut-off levels: 0.81% v/w, 0.79% v/w and 0.82% v/w, respectively. The results showed that the different cut-off levels do not influence the oil yields. Results revealed that the three cut-of levels applied to the cultivated *O. bastetanum* did not affect significantly the oil composition. From an examination, it was suggested that the micro-propagated *O. bastetanum,* cultivated in the area of Lanjarón (Granada, Spain), is a phenolic oregano, because 73.3-78.9% of its oil composition corresponds to phenolic monoterpenes, or to their biogenetic precursors, such as γ- terpinene and p-cymene. In conclusion, due to the constant phenolic content in all cut-off levels, it was considered *O. bastetanum* as an interesting species for medicinal uses, and recommends the 75% cut-off level to obtain the maximum amount of oil per unit area (Pérez-Galindo et al.2005).

The winery-distillery composts, were prepared by mixing different wastes, such as exhausted grape marc, citrus juice waste, tomato soup waste and cattle manure, at the following rates (on a dry weight basis): Compost C1: exhausted grape marc + citrus juice waste [60:40]. Compost C2: exhausted grape marc + tomato soup waste [60:40]. Compost C3: exhausted grape marc + cattle manure [60:40]. The mixtures (about 1800 kg each pile) were composted in a pilot plant, in trapezoidal piles (1.5 m high with a 2 m x 3 m base), by the Rutgers static pile composting system, which maintains a temperature ceiling in the pile, involving on-demand ventilation through temperature feedback control. The bio-oxidative phase of composting was considered finished when the temperature was close to the external value and reheating did not occur. The air-blowing was then stopped to allow the composts to mature over a period of two months. The moisture of the piles was controlled weekly by adding the necessary amount of water to obtain contents above 40%. The final composts displayed a good degree of maturity, as is shown in the following parameters usually used to assess compost maturity: total organic carbon to total nitrogen ratio (TOC/TN) < 20 (16.6 for C1, 16.8 for C2 and 14.5 for C3) (13); cation exchange capacity (CEC) > 60 meq 100 g-1 organic matter (C1 = 139, C2 = 153 and C3 = 163 meq 100 g-1) (14); CEC/TOC > 1.9 meq g-1, 2.54, 2.82 and 2.84 meq g-1 for C1, C2 and C3, respectively (15); water-soluble organic C < 1.5% (1.12% for C1, 1.04% for C2 and 1.25% for C3) (16); and absence of phytotoxicity, according to the germination index (GI) > 50% (73.8% for C1, 83.1% for C2 and 73.1% for C3) (17). *Experimental design:* The different growing media used were prepared by mixing the mature composts (C1, C2 and C3) in increasing proportions (0%, 25% and 50% v/v), using as diluent a traditional mixture of perlite and peat (PP), obtaining a total of seven treatments (substrates). The mixture of perlite and peat was used as control treatment. So, the seven treatments can be summarized as:

i) PP (100%): Control.
ii) C1 (25%) + PP (75%): A.
iii) C1 (50%) + PP (50%): B.
iv) C2 (25%) + PP (75%): C.
v) C2 (50%) + PP (50%): D.
vi) C3 (25%) + PP (75%): E.
vii) C3 (50%) + PP (50%): F.

An aromatic herb, oregano (*O. vulgare*), was used in this study. The experiment was carried out in a greenhouse of the Agricultural Experimental Station of the Instituto Valenciano de Investigaciones Agrarias (IVIA), located in Elche (Alicante, Spain). Commercial seedlings (height 4–5 cm) were planted in pots of 1 L capacity, containing the different substrates, and then were watered. The treatments (growing media) of this experiment were established in a completely randomized plot design with four replicates per treatment. The pots were watered as required and no extra fertilization was applied. At the end of the experiment, on day 126, when the plants reached approximately the commercial transplanting size, they were harvested, washed and dried at

60°C in an air-forced oven for 72 h, after which fresh and dried weights were determined. The volatile compounds were identified in control and compost fertilized oregano samples. Only one of the compost mixtures prepared provided oregano samples with statistically similar total concentration of volatiles to control samples (16.28 ± 1.13 g kg-1), this mixture was F (15.91 ± 0.58 g kg-1). On the other hand, the mixtures B [C1 (50%) + PP (50%)] and C [C2 (25%) + PP (75%)] provided the best results from a volatile point of view, with total concentrations of volatile compounds being 20.85 ± 0.18 g kg-1 and 18.36 ± 1.21 g kg-1, respectively. An increase in the ratios of the C1 (exhausted grape marc + citrus juice waste) from 25% to 50% was related to significant increases in the total concentration of volatile compounds. On the other hand, an increase in the ratio of and C2 (exhausted grape marc + tomato soup waste) and C3 (exhausted grape marc + cattle manure) resulted in significant decreases in the total concentrations of volatiles. A similar situation was found for the yields of oregano shoots, with an increase in C1 causing an increase in fresh yields (47.4 g and 55.9 g plant-1) while an increase in C3 resulted in a decrease in the fresh plant yield from 57.8 g to 37.7 g plant-1; the yield of control plants 33.8 g plant-1 (Bolechowski et al.2011).

A number of relevant studies including the effects of temperature and day length have been conducted by various authors (Cohen et al., 1980; Putievsky, 1983; Dudai, 1988; Putievsky et al., 1988; Dudai et al., 1989; Dudai et al., 1994). According to these studies, air temperature for optimum oregano growth and development range from 18 °C to 22 °C, whereas the root system of well-developed plants (older than one year) may withstand air temperatures from -25 °C to 42 °C (Koutsos, 2006). However, temperatures below 4 °C or above 33 °C may limit plant growth. The plant grows in a wide variety of soils and climates from seaside to mountainous areas on the islands and mainland of Greece (1500 m), in rich as well as on poor calcareous soils. An excellent soil pH value is around 6.8, but oregano may be found on calcareous soils with much higher pH values. Generally, a long light period is needed (more than 12 hours) for a high content of essential oil and carvacrol.

A field experiment was conducted Chauhan et al. 2013 during 2005-2009 to evaluate the 18 accessions of Indian Oregano (*O. vulgare* L.) collected from western Himalaya region of Uttarakhand. All the accessions collected were screened for herb yield and essential oil yield. The highest herb yield was recorded in 'OV7' (926.40 gm) by followed by 'OV4' and 'OV9' with 671.40 and 635.00 gm, respectively. The essential oil yield was found to be highest in accession 'OV7' (1.83 ml). The 'OV4' accession dominated by thymol as the major compound yielded 1.40 ml followed by 'OV9' with carvacrol as major compound yielding 1.26 ml. The result of the experiment evaluated accession 'OV7' for high herb and essential oil yield. While accessions 'OV4' and 'OV9' can be domesticated as high thymol and carvacrol rich lines. The aerial parts of *O. vulgare* were procured from seven wild and three cultivated populations of *O. vulgare* L. collected from diverse microclimatic zones in the Kashmir Himalayas, (Jain et al. 2020). In their report, a variation in essential oil content even among cultivated accessions ranging from 52.99% to 62.05%. Moreover, various compounds in the volatile fraction were not identified in contrast to wild accessions, which had a broader spectrum of constituents. Some chemical constituents, like *p*-cymene, were absent in cultivated accessions but present in wild accessions in very less concentrations. Results showed that 'Greek oregano' plants cultivated under foil were characterized by more numerous and longer shoots in comparison to those grown in open field conditions. The results were reflected in the yield of the herb. Here, plants cultivated under foil were distinguished by almost twice the mass of the raw material when compared to those from the open field. Such results seemed to be associated with high temperature requirements of 'Greek oregano' resulting from its Mediterranean origin.

Another climate-based parameter considered as one of the most important for plant productivity is solar radiation (Boyer, 1982). In his experiment, in the case of foil tunnel conditions, it was recorded at a lower level. However, when comparing the growth of plants cultivated in open field and in the foil tunnel, this parameter did not limit the development of plants from the latter group. It is worth noting that the herb was harvested four times during vegetation season, both in the open field and foil tunnel. Results obtained in this study show that the total content of phenolic acids, flavonoids, and chlorophylls in the 'Greek oregano' herb depends on the used variant of cultivation. In their study, the highest content of phenolic acids was observed in the herb from the first cut. The plants grown in open field were characterized with higher amount of phenolic acids. The accumulation of flavonoids in the herb collected from plants grown in open field was similar to phenolic acids. It was the highest at the first cut, and then, from the second to the last cut, the content was lower but quite uniform, wherein in the cultivation under foil tunnel it decreased significantly from 0.62 (first cut) to 0.20 g×100 g^{-1} (fourth cut). When regard chlorophyll a and b, their content increased from the first to the last cut and was higher in plants cultivated Results of the sensory evaluation indicate that there were no significant differences between the herb from the open field and foil tunnel, in terms of odor and taste attributes Cultivation under foil provides a significantly higher yield of the herb. The raw material from this variant contains more biologically active compounds such as flavonoids, chlorophyll a and b as well as essential oil. Such results seem to be associated with the plant's response to climatic parameters (especially temperature). According to the sensory analysis results, there were no significant differences between the herbs from the open field and foil tunnel, as far as odor and taste attributes are concerned. (Kosakowska et al.2019).

USES

O. vulgare L. is used for the preparation of perfumes, cosmetics and medicines. *Origanum* plants belonging to different species and ecotypes (biotypes) are widely used in agriculture and the pharmaceutical and cosmetic industries as a culinary herb, flavouring substances of food products, alcoholic beverages, and perfumery for their spicy fragrance (Aligiannis et al., 2001, Novak et al.2000; Sivropoulou et al.1996).

Edible Uses

O. vulgare L. is used for the preparation of scented grape wines (Polunin and Stainton, 1984). The above ground parts of this species of *Origanum* are used as a spice in salads, meat dishes, in salting or pickling of vegetables, in non-alcoholic (soft) drinks, and as an insecticide for dried fruit *(*Rybak et al.1989; Hodzhimatov et al. 1971; Hodzhimatov, 1977; Nuraliev, 1986; Bahramov, et al. 1988*)*. In the naturally preserved meat is samarella containing major fraction of oregano (Ulusoy et al. 2018).

Moreover, oregano essential oil has been used as a major food additive in European Union for health benefits (Aguilar et al.2008; Krishnan et al. 2014).

Other Uses

The *O. tytthanthum* oil is used in Russia for soap-making, liquor production, fish canning and perfumery (Rybak et al.1989). It has also been used as a component of inexpensive types of eau-de-Cologne (Pavlov, 1947; Borisova 1954). Oregano has been assessed for its antioxidant and antimicrobial characteristics with particular relevance to food preservation (Chishti et al.2013; Kaul et al. 1996).

Pharmacological Uses

Oregano in Ancient Times

Oregano used to be a single species but has now cultivated around the world into several different varieties. This pungent herb has been cultivated over the last 3,000 years in Egypt. The use of oregano as a disinfectant and preservative was common in Egypt. It was often combined with rosemary and lavender in linen sachets and bath products. It was commonly used for fragrance, cleansing, and hygiene.

In ancient times, oregano was often referred to by the physician Galen and well-known herbalists Culpeper, Parkinson, and Gerard. The *Origanum* species, including marjoram, were recorded for use of a variety of ailments including colds, toothaches, stomachaches, bruises, swellings, earaches, and coughing. Also, in the folk medicine of Middle Asia, *O. tytthanthum* is used to treat gastritis, colitis, bronchitis, and pneumonia. *O. vulgare* L., which belongs to the *Origanum* Section, is widely used as a spice and is an extremely variable species which extends from Macaronesia across a large part of Europe and the Mediterranean to Asia. (Arnold et al. 2000). Oregano (*O. vulgare*) is a native herb of the Mediterranean countries, such as Italy and Spain, and is used as a spicy herb (Figiel et al. 2010). Oregano is of great economic importance due to its use as a spice and as an essential oil. The oregano essential oil is rich in phenolic compounds (e.g., rosmarinic acid, caffeic acid and other hydroxycinnamic acid compounds) and has strong antioxidant, antimicrobial, cytotoxic and antifungal activities (Jerkovic et al. 2001, Wojdyło et al. 2008). Due to all these beneficial effects on human health, oregano essential oil is being used to extend the shelf life of foods (Viuda-Martos et al. 2007). *O. vulgare* plays a primary role among culinary herbs in world trade (Oliver 1997). Despite its economic importance, it is often referred to as an underutilized species, in the sense that its genetic and chemical variability have not yet been fully explored (Padulosi 1997). Traditionally, it is used as a remedy to treat various ailments such as whooping cough, digestive disorders and menstrual problems (Ozbek et al. 2008).

Health Benefits of Oregano

Oregano can be used to kill bacteria, yeast, fungi, parasites, and viruses. Additionally, oregano is effective for intestinal gas, sore throats, coughs, rheumatism, earache, sinusitis, and diarrhea. It is also used to treat nervous tension, breathing difficulties, dandruff, diaper rash, toothache, bee stings, and venomous bites, as well as relieve cramps, reduce fever, and reduce the effects of mumps and measles. *O. onites* is used in traditional medicine. Herbal teas from this plant have been used as therapy for the common cold and stomach ache (Yesilada 1993), abdominal pain, rheumatism (Honda et al. 1996), and as an antiseptic (Kahraman and Kocabas 2001) and antigenotoxic (Ipek et al. 2005).

Oregano's Antimicrobial and Antibacterial Properties

While oregano is most commonly known as a culinary herb in modern times, it contains powerful antimicrobial and antibacterial properties. Studies on oregano essential oil show antibacterial activity, even against *E. coli*. The main active constituents in oregano are carvacrol and thymol, which have broad-spectrum antibacterial action also against *Salmonella typhinium* and *E. coli*, as well as different kinds of MRSA, antibiotic-resistant bacteria.

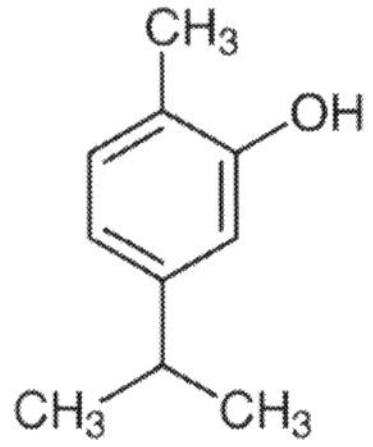

CH_3 OH CH_3 CH_3

Carvacrol

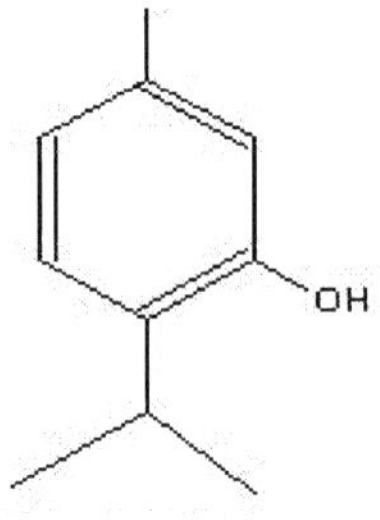

OH

Thymol

Oregano for Arthritis

Its warming properties make it a great analgesic for achy muscles, bones, and joints. The studies show that oregano also contains rosmarinic acid, which was suggested in immunity studies to inhibit T-cell mediated progression of rheumatoid arthritis.

Oregano Lowers Stress

Dietary use of oregano leaf reduced oxidative stress in rats, demonstrating its powerful antioxidant properties as well. Both carvacrol and rosmarinic acid have been reported to protect DNA from damaging agents. In accordance with a wide range of biologically active compounds, this raw material demonstrates various pharmacological activities, including antimicrobial, antioxidant, stimulative, expectorant, carminative, choleretic, antispasmodic, and anticancer. Thus, oregano extracts have been used in both modern and folk medicine for gastrointestinal disorders treatment, temporary loss of appetite, to stimulate bile secretion, and externally in skin and mucous membrane inflammations (Baricevic and Bartol, 2002; Chishti et al. 2013). Comprehensive usage of the 'Greek oregano' herb, both as a spice and as medicine, contributes to its significant economic importance. Nowadays, it is regarded as one of the most traded and consumed spicy/medicinal plant, used in large amounts especially in the USA.

Oregano for Anticarcinogenic Activity

O. syriacum L. and *O. vulgare* L., were analyzed for their antiproliferative activity against breast adenocarcinoma MCF7 (Al-Kalaldeh et al. 2010). The ethanol plant extracts of *O. triloba* exhibited an antiproliferative activity against MCF7 as shown by the IC50 values of 6.40, 24.49, and 25.25 μg/mL. The anti-cancer effects of lyophilized oregano were analyzed in an *in vitro* and *in vivo* breast cancer mice model which showed suppression of tumor frequency by 55.5%, tumor volume by 44.5%, and tumor incidence by 44% compared to control animals (Kubatka et al. 2017).

Origano as Antioxidants

Oregano is commonly used as safe feed additive containing massive phenolic contents with antioxidative characters. Additionally, *O. vulgare* could protect against lung injury and hepatic toxicity. The protective activity of *O. vulgare* against the carbon tetra chloride induced hepatotoxicity was reported by Sikander et al. (2013). Sankar et al. (2014) developed titanium dioxide nanoparticles from *O. vulgare* and used it in wound treatment. They reported a marked wound healing by titanium dioxide nanoparticles of *O. vulgare*. Ethanol extracts (stem and root) of *O. majorana* showed strong antioxidant activity, reducing power ability, free radical scavenging activity and meal chelating ability when compared to standards such as ascorbic acid which may be due to presence of flavonids, phenols, tannins (phenolic compounds) and triterpenoids (Vegi et al. 2005).

Oregano for Anti-Inflammatory Activity

The anti-inflammatory activity of *O. vulgare* was studied by Javadian et al. (2016). By using methanolic leaf extract at a dose of (1.5, 2.25 and 2.7 mg/ml). Leaf extract showed anti-inflammatory effects on the activated mixed and microglial cells through inhibition of iNOS (Javadian et al. 2016).

Table 2. Some pharmacological properties found in *Origanum* species are summarized as follow

S. No.	Plant species	Country	Oil/ Extract	Plant part	Activity	References
1	*O. vulgare* ssp. *hirtum*,	Italy	essential oil	dry leaves	antimicrobial	Tommasi et al. 2009
2	*O. vulgare*	Albania	essential oil	flowering top	antimicrobial	Höferl et al. 2009
3	*O. minutiflorum*	Turkey	essential oil	aerial part during flowering stage	antimicrobial	Altundag et al. 2011
4	*O. vulgare* L.	India	essential oil	aerial part of cultivated and wild ecotypes	antimicrobial	Jain et al. 2020
5	*O. minutiflorum*	Turkey	essential oil	aerial part	antimicrobial	Altundag et al. 2011
6.	*O. vulgare*	Albania	essential oil	flowering top	antimicrobial	Höferl et al. 2009
7	*O. vulgare* ssp. *hirtum*,	Italy	essential oil	leaves	antimicrobial	Tommasi et al. 2009
8	*O.* × *dolichosiphon*	Turkey	essential oil	aerial	antimicrobial	Tabanca et al. 2001
9	*O. vulgare* L.	Jordan and Saudi Arabia	essential oil	leaves and stem	antimicrobial	Khan et al. 2019
10	*O. vulgare*	Algeria	essential oil	aerial part	antimicrobial	Boughendjioua and Seridi, 2017
11	*O. vulgare*	Greece	essential oil	stem and leaves	antimicrobial	Alexopoulos et al. 2017
12	*O. vulgare*	Algeria	essential oil	-	antibacterial	Mahfouf 2017
13	*O. vulgare* ssp. *Virens*	Portugal	essential oil	aerial part in flowering stage	antifungal	Vale-Silva et al. 2012
14	*O. vulgare*	Italy	essential oil		antifungal	Giamperi et al. 2002
15	*O. hypericifolium*	Turkey	essential oil	flowers and fruits	antifungal	Ocak et al. 2012
16	*O. heracleoticum*	Serbia	essential oil	-	antifungal	Dzamic et al. 2008
17	*O. vulgare, T. serpyllum*,	Poland	essential oil	commercially	antifungal	Baj et al. 2020

S. No.	Plant species	Country	Oil/ Extract	Plant part	Activity	References
	T. vulgaris L.			purchased		
18	*O. syriacum*		essential oil	vegetative tissues	bioactive inhibit growth of *C. albicans*	Kassaify et al. 2008
19	*O. vulgare*	Italy	essential oil	whole plant	antimicrobial, antifungal and antioxidant	Baratta et al. 1998
20	*O. vulgare*	Italy	essential oil	-	antimicrobial and antioxidant	Baratta et al. 1998
21	*O. hypericifolium*	Turkey	essential oil	aerial part	antimicrobial and antioxidant	Celik et al. 2010
22	*O. vulgare* ssp. *hirtum*,	Germany	essential oil	aerial part	antioxidant	Dorman and Deans 2004
23	*T. pallescens*	Algeria	essential oil	aerial part	antioxidant	Hazzit et al. 2013
24	*O. vulgare*	Iran	essential oil	aerial part	antioxidant	Roofchae et al. 2011
25	*O. vulgare* ssp. *Gracile*	Iran	essential oil	oven dried	antioxidant	Morshedloo et al. 2017
26	*O. majorana*	Turkey	essential oil	aerial part	antiproliferative, antioxidant	Erenler et al. 2016
27	*O. majorana*	Saudi Arab	essential oil	aerial part	antiulcer antisecretory antioxidant.	Al-Howiriny et al. 2009
28	*O. onites, O. minutiflorum.*	Turkey	essential oil	aerial part	larvicidal	Cetin and Yanikoglu 2006
29	*O. vulgare*	France	essential oil	whole plant	antimalarial	Milahu et al. 1997
30	*O. vulgare*	France	essential oil		antimalarial	Milhau et al. 1997
31	*O. ehrenbergii* Boiss, *O. syriacum* *O. vulgare*	Maxico	essential oil	-	anti-inflammatory	Salud et al. 2011

Table 2. (Continued)

S. No.	Plant species	Country	Oil/ Extract	Plant part	Activity	References
32	*O. vulgare* ssp. *vulgare* *O. vulgare* ssp. *hirtum*	Greece	essential oil	-	did not produce any marked change in the snails' behavior significant repellent effect	Vokou et al. 1998
33	*O. vulgare*	Greece	essential oil	flowers	anti-yeast (Y. lipolytica LFMB 20)	Chatzifragkou et al. 2011
34	*O. onites*	Turkey	essential oil	foliage	inhibitory effect against weed seeds	Azirak and Karaman 2008
35	*Origanum* sp	Yemen	essential oil	leaves and branches	antidiabetic	Ya'ni et al. 2018
36	*O. majorana*	India	essential oil	leaves	anti-diabetic, anti-hyperlipidemic	Tripathy et al. 2018
37	*O. majorana*	Egypt	essential oil	leaves	antimutagenic effect	Mossa et al. 2013
38	*O. vulgare*	India	essential oil	Whole plant	insecticidal	Tandon et al., 2008
39	*O. vulgare*	Thialand	essential oil	whole plant	anti-skin-ageing	Laothaweerungsawat et al. 2020
40	*O. vulgare* ssp. *glandulosum*	Algeria	essential oil and methanolic extract	aerial part during flowering stage	antimicrobial and antioxidant	Bouhaddouda et al. 2016
41	*O. vulgare* ssp. *vulgare*	Romania	extract	-	antioxidant; antimicrobial; hepatoprotective	Oniga et al. 2018
42	*O. vulgare*	Brazil	ethanolic extract	aerial part	antioxidant	Aranha and Jorge 2012
43	*O. vulgare*	Romania	phenolic extract		antioxidant	Spiridon et al. 2011
44	*O. majorana*	India	alkaloid extract	leaves	ntimicrobial	Thiripurasundari et al. 2013

S. No.	Plant species	Country	Oil/ Extract	Plant part	Activity	References
45	***O. majoranum***	Saudi Arab	aqueous extract	leaves	hepatic and renal histopathological	Soliman et al. 2016
46	*O. vulgare*	Romania	ethanolic extract	leaves and flowering stem	antioxidant	Benedec et al. 2018
47	*O. vulgare*	Iran	hydroalcoholic extract	-	inhibitory	Bahmani et al. 2019
48	*O. vulgare*	Iraq	hot aqueous extract	-	antioxidant	Ghani et al. 2020
49	*O. vulgare*	India	water distillation methanolic and ethanolic extract	shoot part leaves	antioxidant	Jan et al. 2020
50	*O. vulgare*	Canada	ethanolic extract	leaves	antioxidant, antibacterial, cytotoxic	Coccimiglio et al. 2016
51	*O. vulgare*	Peru	aqueous extract	leaves	embryotoxicity	Benavides et al. 2011

References

Aguilar, F., Autrup, H., Barlow, S., Castle, L., Crebelli, R. and Dekrant, W. (2008). Use of rosemary extracts as a food additive-scientific opinion of the panel on food additives, flavourings, processing aids and materials in contact with food. *EFSA Journal*, 6(6): 721.

Ai Bandak, G. and Oreopoulou, V. (2007). Antioxidant properties and composition of *Majorana syriaca* extracts. *Euro. J. Lipid Sci. Technol.,*109:247-255.

Alexopoulos, A., Plessas, S., Kimbaris, A., Varvatou, M., Mantzourani, I., Fournomiti,M., Tzouti, V., Nerantzaki, A. and Bezirtzoglou, E. (2017). *Current Research in Nutrition and Food Science*, 05(2): 109-115.

Al-Howiriny, T., Alsheikh, A., Alqasoumi, S., Al-Yahya, M., El-Tahir, K. and Rafatullah, S. (2009). Protective Effect of *Origanum majorana* L. 'Marjoram' on Various Models of Gastric Mucosal Injury in Rats. *The American J. of Chinese Med.*, 37(3), 531–545.

Aligiannis, N., Kalpoutzakis, E., Mitaku, S., Chinou, I.B. (2001). Composition and antimicrobial activity of the essential oils of two *Origanum* species. *J. Agri. Food Chem*, 49: 4168-4170.

Al-Kalaldeh J.Z., Abu-Dahab R. and Afifi F.U. (2010). Volatile oil composition and anti-proliferative activity of *Laurus nobili*s, *Origanum syriacum*, *Origanum vulgare*, and *Salvia triloba* against human breast adenocarcinoma cells. *Nut. Res*., 30: 271–278.

Altundag, S., Aslim, B. and Ozturk, S. (2011). In vitro Antimicrobial Activities of Essential Oils from *Origanum minutiflorum* and *Sideritis erytrantha* subsp. erytrantha on Phytopathogenic Bacteria. *J. Essent. Oil Res*., 23(1): 4-8.

Antuono, L.F.D', Galetti, G.C. and Bocchini, P. (2000). Variability of essential oil content and composition of *Origanum vulgare* L. populations from a north Mediterranean area (Liguria region, northern Italy). *Annal. Bot*., 86:471-478.

Aranha, C.P.M. and Jorge, N. (2012) Antioxidant potential of oregano extract (*Origanum vulgare* L.). *British Food Journal*, 114 (7): 954-965.

Arnold, N., Bellomaria, B. and Valentini, G. (2000). Composition of the Essential Oil of Three Different Species of *Origanum* in the Eastern Mediterranean. *J. Essent. Oil Res.*, 12(2): 192-196.

Arnold, N., Bellomaria, B., Valentini, G. and Arnold, H.J. (1993). Comparative study of the essential oil from three species of *Origanum* growing wild in the Eastern Mediterranean region. *J. Essent. Oil Res.*, 5, 71-77.

Azirak, S. and Karaman, S. (2008). Allelopathic effect of some essential oils and components on germination of weed species Acta Agriculturae Scandinavica. Section B _ *Soil and Plant Science*, 58: 88-92.

Bahmani,M., Taherikalani, M., Khaksarian, M., Rafieian-Kopaei, M., Ashrafi, B., Nazer, Setareh Soroush, M., Abbasi, N. and Rashidipour, M. (2019). T*he synergistic effect of hydroalcoholic extracts of Origanum vulgare, Hypericum perforatum and their active components carvacrol and hypericin against Staphylococcus aureus,* 5(3):FSO371.

Bahramov, A.B., Dzhumaev, N., Salihov, S.A. and Devlikamova, T.M. (1998). *Ziziphora sp.* and *Origanum tyttbanthum* for tinned (preserve) cucumbers. *Pischevaja promischlennost*, 7, 45.

Baj, T., Biernasiuk, A., Wróbel, R. and Malm, A. (2020) Chemical composition and *in vitro* activity of *Origanum vulgare* L., *Satureja hortensis* L., *Thymus serpyllum* L. and *Thymus vulgaris* L. essential oils towards oral isolates of *Candida albicans* and *Candida glabrata*. *Open Chem.*, 18: 108–118.

Baratta, M.T., Dorman, H.J. D., Deans, S.G., Biondi, D.M. and Ruberto, G. (1998). Chemical composition, antimicrobial and antioxidative activity of Laurel, Sage, Rosemary, Oregano and Coriander essential oils. *J. Essent. Oil Res.*, 10(6): 618-627.

Baricevic, D. and Bartol, T. (2002). The biological/pharmacological activity of the *Origanum* genus. In: Kintzios SE (ed) *Medicinal and aromatic plants-Industrial profiles, Oregano.* The genera *Origanum* and *Lippia*, Taylor and Francis, London, 25:177–213.

Baser, K.H.C., Kirimer, N. and Tümen, G. (1993). Composition of the essential oil of *Origanum majorana* L. from Turkey. *J. Essent. Oil Res.,* 5: 577-579.

Baser, K.H.C., Özek, T., Kürkcüoglu, M. and Tümen, G. (1994). *The essential oil of Origanum vulgare subsp. hirtum of Turkish origin. J. Essent. Oil Res*., 6, 31-36.

Benavides, V., D'Arrigo, G. and Pino, J. (2011). Effects of aqueous extract of *Origanum vulgare* L. (Lamiaceae) on the preimplantational mouse embryos. *Rev. peru. biol.* 17(3): 381 – 384.

Benedec, D., Oniga, I., Cuibus, F., Sevastre, B., Stiufiuc, G., Duma, M., Hanganu, D., Iacovita, C., Stiufiuc, R. and Lucaciu, C.M. (2018). *Origanum vulgare* mediated green synthesis of biocompatible gold nanoparticles simultaneously possessing plasmonic, antioxidant and antimicrobial properties, *Int. Jr. of Nanomedicine*, 13: 1041–1058.

Bisht, D., Pal, A., Chanotiya, C.S., Mishra D. and Pandey, K.N. (2011). Terpenoid composition and antifungal activity of three commercially important essential oils against *Aspergillus flavus* and *Aspergillus niger*. *Nat. Prod. Research: Formerly Natural Product Letters,* 25:20, 1993-1998.

Bolechowski, A., Moral, R., Bustamante, M.A., Paredes, C., Agulló, E., Bartual, J. and Carbonell-Barrachina, A.A. (2011): Composition of oregano essential oil (*Origanum vulgare*) as affected by the use of winery-distillery composts, *J. Essent. Oil Res*., 23(3): 32-38.

Borisova, A.G. (1954). The genus *Origanum.* In the "Flora USSR." Moscow-Leningrad.

Bosabalidis, A. and Tsekos, I. (1982). Glandular scale development and essential Oil secretion in *Origanum dictamnus* L. *Planta*, 156: 496-504.

Bosabilidis, A.M. and Tsekos, I. (1984). Glandular hair formation in *Origanum* Species. *J. Bot*., 53:559-563.

Boughendjioua, H. and Seridi, R. (2017) Antimicrobial efficacy of the essential oil of *Origanum Vulgare* from Algeria. *J Pharm Pharmacol Res,* 1 (1): 019-027.

Bouhaddouda N., Aouadi S. and Labiod R. (2016). Evaluation of chemical composition and biological activities of essential Oil and methanolic extract of *Origanum vulgare* L. *ssp. glandulosum* (Desf.) *Ietswaart* from Algeria, *Int. Jr. of Pharmacognosy and Phytochemical Research*, 8(1):104-112.

Celik, A., Herken, E.N., Arslan, I., Ozel, M.Z. and Mercan, N. (2010). Screening of the constituents, antimicrobial and antioxidant activity of endemic *Origanum hypericifolium* O. Schwartz & P.H. Davis, *Nat. Prod. Research*: *Formerly Natural Product Letters,* 24(16): 1568-1577.

Chalchat, J.C., Pasquier, B. (1998). Morphological and chemical studies of *Origanum* clones*: Origanum vulgare* L.ssp. *vulgare. J. Essent. Oil Res*., 10:119-125.

Chatzifragkou, A., Petrou, I., Gardeli, C., Komaitis, M., Papanikolaou, S. (2011) Effect of *Origanum vulgare* L. essential oil on growth and lipid profile of *Yarrowia lipolytica* cultivated on glycerol-based media. *J Am Oil Chem Soc.*, 88:1955–1964.

Chishti, S., Kaloo, Z.A. and Sultan, P. (2013). Medicinal importance of genus *Origanum*: a review. *J. Pharmacognosy Phytother.* 5 (10): 170–177.

Coccimiglio, J., Alipour, M., Jiang, Z.H., Gottardo, C. and Suntres, Z. (2016). A*ntioxidant, antibacterial, and cytotoxic activities of the ethanolic Origanum vulgare extract and its major constituents, oxidative medicine and cellular longevity,* dx. doi. org/ 10.1155/2016/ 1404505.

Cohen, A., Putievsky, E., Dafni, A., Fleisher, A. (1980). Seed germination of wild spices from the "oregano" type. *Hassadeh*, 60:1160-1162.

Davis, P.H. (1982). *Flora of Turkey and the East Aeg~n~sland* Crambridge University Press, Cambridge, 3:171.

Deans, S.G. and Ritchie, G. (1987). Antibacterial properties of plant essential oils. *Int. J. Food Microbiol*., 5:165-180.

Dordas, C. (2012). *Aromatic and medicinal plants.* Modern Education (Eds), Thessaloniki.

Dorman, H.J. D. and Deans, S.G. (2004). Chemical composition, antimicrobial and *In vitro* antioxidant properties of *Monarda citriodora* var. *citriodora*, *Myristica fragrans*, *Origanum vulgare* ssp. *hirtum*, *Pelargonium* sp. and *Thymus zygis* oils, *Journal of Essential Oil Research,* 16(2): 145-150.

Dorman, H.J.D. and Deans, S.G. (2004). Antimicrobial agents from plants: antibacterial activity of plant volatile oils. *J. Applied Microbiol.*, 88, 308-316.

Dudai, N. (1988). *Environmental factors affecting flowering, morphology and essential oil of Origanum syriacum var 'syriacum'.* MSc Thesis, Hebrew University of Jerusalem.

Dudai, N., Putievsky, E., Palevitch, D., Halevy, A.H. (1989). Environmental factors affecting flower initiation and development in *Majorana syriaca* L. (*Origanum syriacum* var. *syriacum*). *Israel J. of Botany*, 38:229- 239.

Dudai, N., Putievsky, E., Ravid, U., Palevitch, D., Halevy, A.H. (1992). Monoterpene content in *Origanum syriacum* as affected by environmental conditions and flowering. *Physiologia Plantarum*, 84:453-459.

Duke JA. (1992). *Dr. Duke's phytochemical and ethnobotanical databases,* http://www.ars-grin.gov/duke/

Dzamic, A., Sokovic, M., Ristic, M.S., Grujic-Jovanovic, S., Vukojevic, J. and Marin, P.D. (2008). Chemical composition and antifungal activity of *Origanum heracleoticum* essential oil. *Chemistry of Natural Compounds,* 44(5).

Elgayyar, M., Draughon, F.A., Golden, D.A. and Mount, J.R. (2001). Antimicrobial activity of essential oils from plants against selected pathogenic and *saprophytic* microorganisms. *J. Food Protect.,* 64:1019-1024.

Erenler, R., Sen, O., Aksit, H., Demirtas, I., Yaglioglu, A.S., Elmastasa, M. and Telci, I. (2016) Isolation and identification of chemical constituents from *Origanum majorana* and investigation of antiproliferative and antioxidant activities. *J. Sci. Food Agric.* 96: 822–836.

Figiel, A., Szumny, A., Gutiérrez-Ortíz, A. and Carbonell-Barrachina, A.A. (2010). Composition of oregano essential oil (*Origanum vulgare*) as affected by drying method. *J. Food Eng.*, 98: 240-247.

Ghani J.M., Sharba I.R., Shamran S.J. (2020). Antioxidant role of *Origanum vulgare* extract against reproductive toxicity assessment of fenugreek in female mice, *Eur Asian J. of Bio Sciences*, 14:1335-1340.

Giamperi, L., Fraternale, D. and Ricci, D. (2002). The *In vitro* action of essential oils on different organisms, *J. Essent. Oil Res.*, 14(4): 312-318.

Goliaris, A. (1997). *Origano cultivation in unproductive Greek land. Scientific Bulletin,* Agricultural Resource Center of Northern Greece, 4:79-86.

Goliaris, A.H., Chatzopoulou, P.S., and Katsiotis, S.T. (2002). Production of new Greek Origano clones and analysis of their essential oils. *J. of Herbs Spices and Medicinal Plants, 10:29*-35.

Graven, E.H., Deans, S.G., Svoboda, K.P., Mavi, S. and Gundiza, M.G. (1992). Antimicrobial and antioxidative properties of the volatile (essential) oil of Artemisia afra Jacq. *Flav. Fragr. J.*, 7:121-125.

Grevsen, K., Fretté, X. and Christensen, L.P. (2009). Content and composition of volatile terpenes, flavonoids and phenolic acids in Greek oregano (*Origanum vulgare* L. Ssp. *hirtum*) at different development stages during cultivation in cool temperate climate. *Eur. J. Hortic. Sci.*, 74 (5): 193–203.

Hazzit, M., Baaliouamer, A. and Douar-Latreche, S. (2013). Effect of heat treatment on the chemical composition and the antioxidant activity of essential oil of *Thymus pallescens* de Noé from Algeria, *J. Essent. Oil Res.*, 25(4): 308-314.

Hedge, L. (1990). Labiatae. In S.I. Ali and Y.I. Nasir (Eds.), *Flora of Pakistan* (Labiatae) Karachi: University of Karachi, 192: 244–247.

Hodzhimatov, K.C. (1977). *Instruction about Collect, Dry and Save Spice-Aromatic Plants.* Information report, N 179, Taschkent.

Hodzhimatov, K.C., Sagatov, S.S., Haitmuhamedov, L.P. (1971). The dynamic of accumulation essential oil of *Origanum tytthanthum. Uzbek. Biol. J.,* 6: 12-13.

Hoferl, M., Buchbauer, G., Jirovetz, L., Schmidt, E., Stoyanova, A., Denkova, Z., Slavchev, A. and Geissler. (2009). Correlation of antimicrobial activities of various essential oils and their main aromatic volatile constituents, *J. Essent. Oil Res.*, 21(5), 459-463.

Honda, G., YeÕilada, E., Tabata, M., Sezik, E., Fujita, T., Takeda, Y., Takaishi, Y. and Tanaka, T. (1996). Traditional medicine in Turkey VI. Folk medicine in West Anatolia: Afyon, Kütahya, Denizli, Mu—la, Aydin provinces. *J. Ethnopharm*. 53: 75-87.

Huseyin, C. and Yanikoglu, A. (2006). A study of the larvicidal activity of *Origanum* (Labiatae) species from southwest Turkey. *Journal of Vector Ecology* 31 (1): 118-122.

Ietswaart, J.H. (1980). *A Taconomic Revision of the Genus Origanum (Labiatae).* Ph.D. thesis. Leiden Botanical Series, 4, The Hague:Leiden University Press.

Ietswaart, J.H. (1982). *Origanum* L., In: *Flora of Turkey and the East Aegean Islands.* Edit., P.H. Davis, University Press, Edinburgh, 7:308.

Ipek, E., Zeytinoglu, H., Okay, S., Tuylu, B.A., Kurkcuoglu, M. and Baser. K.H.C. (2005). Genotoxicity and antigenotoxicity of *Origanum* oil and carvacrol evaluated by Ames Salmonella/microsomal test, *Food Chem.* 93: 551-556.

Ivask, K., Orav, A., Kailas, T., Raal, A., Arak, E. and Paaver, U. (2005). Composition of the essential oil from wild marjoram (*Origanum vulgare* L.ssp. *vulgare*) cultivated in Estonia, *J. Essent. Oil Res.*, 17(4):384-387.

Jan, S., Rashid, M., Abd_Allah, E.F. and Ahmad, P. (2020). Biological efficacy of essential oils and plant extracts of cultivated and wild ecotypes of *Origanum vulgare* L., *Bio Med Research Int.*, doi.org/10.1155/2020/8751718.

Javadian, S., Sabouni, F. and Haghbeen, K. (2016). *Origanum Vulgare* L. extracts versus thymol: An anti-inflammatory study on activated microglial and mixed glial cells. *J. of Food Biochemistry*. 40(1):100-8.

Jerkovic, I., Mastelic, J. and Miloš, M. (2001). The impact of both the season of collection and drying on the volatile constituents of

Origanum vulgare L. ssp. *hirtum* grown wild in Croatia. *Int. J. Food Sci. Technol.*, 36: 649-654.

Joshi PK, Tandon S, Pant AK and Mathela CS. (2011). Chemical composition and antifungal activity of essential oil liverwort *Cryptomitrium himalyense* Kash. *Indian Perfumer,* 55(3): 33-34

Joshi, P.K., Joshi, N., Tewari, G. and Tandon, S. (2016). Chemical, biocidal and pharmacological aspects of *Origanum* Species (A brief Review), *J. of Indian Chemical Society*, 92:1605-1613.

Kahraman, S. and Kocabas. Y.Z. (2001). Traditional medicinal plants of Kahraman Maras (Turkey). *The Sciences*. 1: 125-128.

Kassaify, Z., Gerges, D.D., Jaber, L.S., Hamadeh, S.K., Saliba, N.A., Talhouk, S.N. and Barbour, E.K. (2008). Bioactivity of *Origanum syriacum* essential oil against *Candida albicans*, *J. of Herbs, Spices & Medicinal Plants*, 14(3-4), 185-199.

Katsiotis. S. and Chatzopoulou. P. (2010). *Aromatic, medicinal and essential oil.* Adelfhon Kyriakidi Ed, Thessaloniki.

Kaul, V.K., Singh, B. and Sood, R.P. (1996). "Essential oil of *Origanum vulgare* L. from north India, *J. Essent. Oil Res*., 8(1), 101–103.

Khan, M., Khan, S.T., Khan, M., Mousa, A.A., Mahmood, A. and Alkhathlan, H.Z. (2019). *Chemical diversity in leaf and stem essential oils of Origanum vulgare L. and their effects on microbicidal activitie*s, AMB express, 9, Article no. 176.

Kokkini, S. and Vokou, D. (1989). Carvacrol rich plants in Greece. *Flav. Fragr. J.*, 4, 1-7.

Kokkini, S., Karousou, R., Dardioti, A., Krigas, N. and Lanaras, T. (1997). Autumn essential oils of Greek oregano. *Phytochemistry*, 44: 883-886.

Kosakowska, O., Węglarz, Z. and Bączek, K., (2019). Yield and quality of 'Greek oregano' (*Origanum vulgare* L. subsp. *hirtum*) herb from organic production system in temperate climate, *Ind. Crops & Products*, Vol 141, DOI: 10.1016/j.indcrop.2019.111782.

Koutsos, T. (2006). *Aromatic and medicinal plants.* Ziti, Thessaloniki, Greece.

Kubatka P., Kello M., Kajo K., Kruzliak P., Vybohova D., Mojzis J., Adamkov M., Fialova S., Veizerova L. and Zulli A. (2017). Oregano demonstrates distinct tumour-suppressive effects in the breast carcinoma model. *Europ. J. Nutr.*, 56: 1303–1316.

Krishnan, R.K., Babuskin, S., Babu, P.A.S. (2014). Bio protection and preservation of raw beef meat using pungent aromatic plant substances. *J. of the Sc. of Food and Agric.*, 94(12): 2456–2463.

Lagouri, V., Blekas, G., Tsimidou, M., Kokkini, S. and Boskou, D. (1993). Composition and antioxidant activity of essential oils from Oregano plants grown wild in Greece. *Zeit. Lebensmit. Untersuch. Forsch.*, 197, 20-23.

Laothaweerungsawat, N., Sirithunyalug, J. and Chaiyana, W. (2020). Chemical compositions and anti-skin-ageing activities of *Origanum vulgare* L. essential oil from Tropical and Mediterranean region. *Molecules*, 25, 1101-15.

Lawrence, B.M. (1993). Spanish oregano oil. *Perfum. Flavor., lwl*, 53-54.

Lawrence, B.M. (1984). The botanical and chemical aspects of Oregano. *Perfum. Flavor., 9*(5):41-51.

Lawrence, B.M. (1989). Progress in essential oils: *Origanum* oil (Greek-type). *Perfum. Flavor.*, 14(1): 36-40.

Lawrence, B.M. (1995). Progress in essential oils: *Origanum oil* - Greek type. *Perfum. Flavor.*, 20(5): 98-102.

Maarse, H. and Van Os, F.H.L. (1973). Volatile oil of *Origanum vulgare* L. ssp. *vulgare*. Oil content and quantitative composition of the oil. *Flav. Indust.*, 4, 481-484.

Mahfouf, N. (2017) Antibacterial activity of the essential oil of *Origanum vulgare* L. against strains of *Acinetobacter spp*. 18th global Summit on Food & Beverages October 02-04, 2017, *J. Food Process Technol.* Chicago, USA DOI: 10.4172/2157-7110-C1-066.

Makinen SM and Pa¨a¨kko¨nen KK (2002) Processing, effects and use of oregano and marjoram in foodstuffs and in food preparation. In: S.E. Kintzios, *Medicinal and aromatic plants-Industrial profiles, Oregano.* The genera *Origanum* and *Lippia*, Taylor and Francis, London, pp 25: 217–233.

Marzi, V. (1996). Agricultural practices for oregano. In: Padulosi A (Ed). *Proceedings of the IPGRI International Workshop on Oregano* 8-12 May 1996, CIHEAM, Valenzano, Bari, Italy.

Melegari, M., Severi, F., Bertoldi, M., Benvenuti, S., Circetta, G., Morone Fortunato, I., Bianchi, A., Leto, C. and Carrubba, A. (1995). Chemical characterization of essential oils of some *Origanum vulgare* L. subspecies of various origin. *Rivista Ital. EPPOS*, 7: 21-28.

Milhau, G., Valentin, A., Benoit, F., Mallié, M., Bastide, J.M., Pélissier, Y. and Bessière, J.M. (1997). *Invitro* antimalarial activity of eight essential oils, *J. Essent. Oil Res.*, 9(3): 329-333.

Mockute, D., Bernotiene, G. and Judzentiene, A. (2001). The essential oil of *Origanum vulgare* L. ssp. *vulgare* growing wild in Vilnius district (Lithuania). *Phytochemistry*, 57, 65-69.

Morshedloo, M.R., Mumivand, H., Craker, L.E. and Maggi, F. (2017). Chemical composition and antioxidant activity of essential oils in *Origanum vulgare* subsp. gracile at different phonological stages and plant parts. *J. of Food Processing and Preservation.* Doi:Org/ 10.1111/jfpp.13516.

Mossa, A.T.H., Refaie, A.A., Ramadan, A. and Bouajila, J. (2013) Antimutagenic effect of *Origanum majorana* L. essential oil against prallethrin-induced genotoxic damage in rat bone marrow cells. *J. of Medicinal Food,* 16 (10): 1–7.

Mouterde, P. (1984). *Nouvelle Flore du Liban et de la Syrie. Dar El Machreq Sarl Beyrourh*, 3: 184.

Mukerjee, S.K. (1940). A revision of the Labiatae of the Indian empire. *Records of Botanical Survey of India,* 14:94–95.

Nakatani, N. (1994). Antioxidant and antimicrobial constituents of herbs and spices. In: Spices, Herbs and Edible Fungi. *Developments in Food Science* 34. Edit., G. Charalambous, Elsevier, London, 251-271.

Novak, J., Bitsch, C.J., Langbehn, F., Pank, M., Skoula, Y., Gotsiou (2000). Ratios of *cis*and *trans*-sabinene hydrates in *Origanum majorana* L. and *Origanum microphyllum* (Benth.) *Vogel. Biochem. Syst. Ecol.* 28: 697-704.

Novak, J., Lukas, B. and Franz, C. (2010). Temperature Influences thymol and carvacrol differentially in *Origanum* spp. (Lamiaceae), *J. Essent. Oil Res*., 22(5): 412-415.

Nuraliev, Y.M., Safarov, N.M., Kurbanov, M.K., Dzhuraev, K.S. and Davlatkadamov, S.M. (1986). (*O. tytthanthum Gontsch – apenpective substitute for O. vulgare L.* raw material in Tadjikistan. *Rast. Resursy*, 22:337.

Nykänen, I. (1986). High resolution gas chromatographic-mass spectrometric determination of the flavour composition of wild marjoram *(Origanum vulgare L.)* cultivated in Finland. Z. Lebensm. *Unters Forsch*, 183: 267-272.

Ocak, I., Çelik, A., Ozel, M.Z., Korcan, E. and Konuk, M. (2012). Antifungal activity and chemical composition of essential oil of *Origanum hypericifolium, Int. J. of Food Properties,* 15(1): 38-48.

Oliver, G.W. (1997). The world market of Oregano. In: S. Padulosi eds. *Oregano. Proceedings of the IPGRI International Workshop on Oregano, 8*-12 May, Valenzano (Bari), Italy, Rome, IPGRI. 142-146.

Oniga,I., Puscas, C., Silaghi-Dumitrescu, R. Olah, N.K., Sevastre, B., Marica, R., Marcus, I., Sevastre-Berghian, A.C., Benedec, D., Pop, C.E. and Hanganu, D. (2018). *Origanum vulgare* ssp. *vulgare*: Chemical composition and biological studies. *Molecules*, 23: 2077-2090.

Özbek, T., Güllüce, M., Sahin, F., Özkan, H., Sevsay, S. and Baris, O. (2008). Investigation of the anti-mutagenic potentials of the methanol extract of *Origanum vulgare* L. subsp. *vulgare* in the Eastern Anatolia Region of Turkey. *Turk. J. Biol*. 32(4): 271-276.

Ozel, M.Z., and Kaymaz, H. (2004). Superheated water extraction, steam distillation and soxhlet extraction of essential oils of *Origanum onites*. *Analytical and Bioanalytical Chemistry*, 379: 1127–1133.

Padulosi, S. (1997). *Oregano. Proceedings of the IPGRI International Workshop on Oregano,* 8-12 May 1996, Valenzano (Bari), Italy, Rome: IPGRI.

Pande, C. and Mathela, C.S. (2000). Essential oil composition of *Origanum vulgare* L. ssp. *vulgare* from the Kumaon Himalayas. *J. Essent. Oil Res.,* 2:441-442.

Pankaj S., Kothiyal, P. and Ratan, P. (2018). Pharmacological and phytochemical studies of *Origanum vulgare*: A review, *Int. Research J. of Pharmacy*, 9 (6):30-34.

Paster, N. Menasherov, M., Ravid, U. and Juven, B. (1995). Antifungal activity of oregano and thyme essential oils applied as fumigants against fungi attacking stored grain. *J. Food Protect.*, 58: 81-85.

Pavlov, N.V. (1947). *Plant's Raw Materials of Kazahstan.* Moscow-Leningrad.

Pérez-Galindo, P., Martínez-Raya, A., Francia, J.R., José Jiménez and Concepción Navarro (2005): *Origanum bastetanum* Soc. cultivated Specimens: characterization of its Essential oil, *J. Essent. Oil Res.*, 17:6, 651-655.

Peter, K.V. (2004). Growth habit of wild oregano populations. In: *Handbook of herbs and spices.* 2. Woodhead Publishing.

Polunin, A., and Stainton, A. (1984). *Flowers of the Himalaya.* New Delhi: Oxford University Press.

Putievsky, E. (1983). Temperature and day-length influences on the growth and germination of sweet basil and oregano. *J. of Horticultural Science,* 58:583-587.

Putievsky, E., Ravid, U., Dudai, N. (1988). Phenological and seasonal influences on essential oil of a cultivated clone of *Origanum vulgare* L. *J. of Sc. Food Agric.,* 43:225-228.

Ravid, U. and Putievsky, E. (1986). Cawacrol and thymol chemotypes of east Mediterranean wild labiatae herbs. In: *Progress in essential oil research.* Edit., E.J. Brunke, Walter de Gruyter & Co., Berlin, 163-167.

Roofchae, A., Irani, M., Ebrahimzadeh, M.A. and Akbari, M.R. (2011). Effect of dietary oregano (*Origanum vulgare* L.) essential oil on growth performance, cecal microflora and serum antioxidant activity of broiler chickens, *African J. of Biotech.*, 10(32), 6177-6183.

Rybak, G.M., Romanenko, L.R. and Korableva, O.A. (1989). *Thespices.* Kiev.

Salud, P.G., Miguel, Z.S., Lucina, A.G. and Miguel, R.L. (2011). Anti-inflammatory activity of some essential oils, *J. Essent. Oil Res.*, 23(5): 38-44.

Sankar R., Dhivya R., Shivashangari K.S. and Ravikumar V. (2014). Wound healing activity of *Origanum vulgare* engineered titanium dioxide nanoparticles in Wistar Albino rats. *J. Mat. Sci. Mat. Med.,* 25: 1701–1708.

Sezik, E., Tümen, G., Kirimer, N., Özek, T. and Baser, K.H.C. (1993). Essential oil composition of four *Origanum vulgare* subspecies of Anatolian origin. *J. Essent. Oil Res.,* 5: 425-431.

Sikander M., Malik S., Parveen K., Ahmad M., Yadav D., Hafeez Z.B. and Bansal M. (2013). Hepatoprotective effect of *Origanum vulgare* in Wistar rats against carbon tetrachloride induced hepatotoxicity. *Protoplasma*, 250: 483–493.

Sivropoulou, A., Papanicolaou, E., Nikolaou, C., Kokkini, S., Lanaras, T., Arsenakis, M. (1996). Antimicrobial and cytotoxic activities of *Origanum* essential oils. *J. Agri. Food. Chem.* 44: 1202-1205.

Skoula, M., Gotsiou, P., Naakis, G. and Johnson, C.B. (1999). A chemosystematic investigation on the mono- and sesquiterpenoids in the genus *Origanum* (Labiatae). *Phytochemistry*, 52: 649-657.

Socorro, O., Tárrega, I. and Rivas, F. (1998). Essential oils from wild and micropropagated plants of *Origanum bastetanum*. *Phytochemistry*, 48, 1347-1349.

Soliman, M.M., Nassan, M.A. and Ismail, T.A. (2016). *Origanum* Majoranum extract modulates gene expression, hepatic and renal changes in a rat model of type 2 diabetes, *Iranian J. of Pharmaceutical Research*, 15 (Special issue): 45-54.

Spiridon, I., Colceru, S., Anghel, N., Teaca, C.A., Bodirlau, R. and Armatu, A. (2011). Antioxidant capacity and total phenolic contents of oregano (*Origanum vulgare*), lavender (*Lavandula angustifolia*) and lemon balm (*Melissa officinalis*) from Romania, *Nat. Prod. Research*: Formerly Natural Product Letters, 25 (17): 1657-1661.

Strid, A. and Tan, V. (1991). *Mountain* Flora *of Greece.* Edinburgh, University Press, 2: 136-137.

Tabanca, N., Demirci, F., Ozek, T., Tumen, G., and Baser, K.H. C. (2001) Composition and antimicrobial activity of the essential oil of *Origanum × dolichosiphon* P.H. Davis. *Chemistry of Nat. Compounds*, 37(3):238-241.

Tandon, S., R., Bali and Pant, A.K. (2001). Insecticidal properties of some indigenous plant against *Callosobruchus chinensis* L. *Ind. J. of Plant Protection*, 29 (1&2): 157-158.

Tandon, S., and Mittal, A.K. (2018). Insecticidal and growth inhibitory activity of essential oils of *Boenninghausenia albiflora* and *Teucrium quadrifarium* against *Spilarctia obliqua*. *Biochem. Systematics and Ecology*, 81: 70-73.

Tandon, S., Mittal, A.K. and Pant, A.K. (2009). Growth regulatory activity of *Trichilia connaroides* Wight. and Arn (Syn: *Haynea trijuga* Roxb) leaf extracts against Bihar hairy caterpillar (*Spilosoma obliqua* Walker). *Int. J. of Tropical Insect Science,* 29(4): 180-184.

Tandon, S., Verma, R.C., Joshi, P.K. and Pant, A.K. (2005). Effect of essential oils of *Elsholtzia* spp. against plant pathogenic fungi. *Indian Perfumer,* 49(3): 321-323.

Tandon, S., Mittal, A.K. and Pant A.K. (2008). Evaluation of essential oils from some aromatic plants against pulse beetle (*Callosobruchus chinensis* L.). *J. of Food Legumes,* 21(4): 281-282.

Thiripurasundari, N., Vinodhkumar, T. and Ramanathan, G. (2013) Antimicrobial potential of medicinal plant extracts against human pathogens, *Int. Research J. of Pharmaceutical and Applied Sciences,* 3(4):107-109.

Tommasi, L., Negro, C., Miceli, A. and Mazzotta, F. (2009). Antimicrobial activity of essential oils from aromatic plants grown in the Mediterranean area, *J. Essent. Oil Res*., 21(2): 185-189.

Tripathy, B., Satyanarayana, S., Khan, K.A. and Raja, K. (2018) Evaluation of anti-diabetic and anti-hyperlipidemic activities of ethanolic leaf extract of *Origanum majorana* in streptozotocin induced diabetic rats. *Int. J. of Pharmaceutical Science and Res.,* 9(4): 1529-1536.

Tutin, T.G., Heywood, Y.H., Burges, N.A., Moore, D.M., Valentine, D.H., Walters, S.M. and Webb, D.A. (1972). *Flora Eumgwea.* Cambridge University Press, Cambridge, 3: 171.

Ulusoy, B., Hecer, C., Kaynarca, D. and Berkan, S. (2018). "Effect of oregano essential oil and aqueous oregano infusion application on microbiological properties of samarella (tsamarella), a traditional meat product of Cyprus," *Foods (Basel, Switzerland)*, 7(4):43.

Vagi E., Rapvi E., Hadolin M., Vasarhelyine Perdei, K., Balazs, A., Blazovics, A. and Simandi, B. (2005). Phenolic and triterpenoid antioxidants from *Origanum majorana* L., herb and extracts obtained with different solvents, *J. Agric Food Chem*, 53: 17-21.

Vale-Silva, L., Silva, M.J., Oliveira, D., Gonçalves¸ M.J., Cavaleiro, C., Salgueiro, L. and Pinto, E. (2012). Correlation of the chemical composition of essential oils from *Origanum vulgare* subsp. *virens* with their in vitro activity against pathogenic yeasts and filamentous fungi. *J. of Med. Microbiology*, 61: 252–260.

Viuda-Martos, M., Ruiz-Navajas, Y., Fernández-López, J. and Pérez-Alvárez, J.A. (2007). Antifungal activities of thyme, clove and oregano essential oils. *J. Food Safety*, 27: 91-101.

Vokou, D., Tziolas, M. and Bailey, S.E.R. (1998). Essential-oil-mediated interactions between oregano plants and helicidae Grazers. *J. of Chemical Ecology*, 24(7):1187-1202.

Wojdyło, A., Figiel, A. and Oszmian´ski, J. (2008). Dehydration techniques effect on polyphenols content, antioxidant activity and color of oregano herb. *J. Clin. Biochem. Nutr.*, 43: 1-4.

Yani A.A., Eldahshan O.A., Hassan S.A., Elwan Z.A. and Ibrahim H.M. (2018) Antidiabetic effects of essential oils of some selected medicinal Lamiaceae plants from Yemen against a-Glucosidase enzyme, *J. of Phytochemistry and Biochemistry*, 2:1.

Yildirim, E., Kesdek, M., Aslan, I., Calmasur, O. and Sahin, F. (2005). The effects of essential oils from eight plant species on two pests of stored product insects. *Frese. Environ. Bull*, 14: 23-27.

BIOGRAPHICAL SKETCHES

Prasoon K. Joshi

Affiliation: Higher Education, Uttarakhand State, India

Education: MSc (Organic Chemistry), PhD (Natural Products), U-SET

Research and Professional Experience: 17 Years Research Experience in the field of Natural Products as well as Teaching Experience of Undergraduate and Post Graduate Students

Professional Appointments: Assistant Professor

Publications from the Last 10 Years

Joshi, P.K., Joshi, N. Tewari, g. and tandon, S. (2015). Chemical, Biocidal and Pharmacological Aspects of *Origanum* species: A brief review, *Journal of the Indian Chemical Society*, 92:1603-1615.

Joshi, P.K., Tiwari, A. and Padalia, R.C. (2020). Chemical and bioactive potentials of *Artemisia* species: A brief review literature. Chapter 13, *In: Natural Products and their Utilization Pattern*, Nova Science Publishers, USA, ISBN 978-1-53618-140-1.

Kandpal, V., Joshi, P. K. and Joshi, N. (2016). GC-MS Analysis of seed essential oil of *Chenopodium ambrosiodes* L. collected from Himalayan region, *Journal of Essential Oils bearing Plants*, 19(1): 258-261.

Kandpal, V., Joshi, P.K. and Joshi, N. (2016). Chemical composition of essential oil from seeds of *Anisomeles indica* from Kumaon foothills, India, *Journal of Chemical Engineering and Chemistry Research*, 3(11): 1109-1113.

Kandpal, V., Joshi, P.K. and Joshi, N. (2020). Chemical compositions of essential oils of roots, shoot and inflorescence of *Selinum wallichianum*, *Journal of Essential Oils Bearing Plants*, 23(4): 795-802.

Kandpal. V., Joshi, P.K. and Joshi, N. (2017). Chemical composition of leaf essential oil of *Hyptis suaveolens* (L.) Poit, *Journal of the Indian Chemical Society*, 92(1): 201-203.

Khati, D., Durgapal, A, and Joshi, P. K. (2016). Biofertilization enhances productivity and nutrient uptake of foxtail millet plants, *Journal of Crop improvement*, DOI:0.1080/15427528.2015.1.105343.

Khati, D., Durgapal, A., Joshi, P.K. and Pande, A. (2020). Use of PGPR for the improvement of growth, yield and nutrients of Horsegram, *Annals of Agri-Bio Research*, 25 (1): 67-76.

Rautela, R., Joshi, P. K., Durgapal, D. and Kandpal, V. (2018). Elementary: A text book of Chemistry, A Pragati Edition, Meerut, India. ISBN 978-93-87812-00-0.

Tewari, G., Khati, D., Rana, L., Yadav, P., Pande, C., Bhatt, S., Kumar, V., Joshi, N., and Joshi, P.K. (2016). Assessment of Physicochemical properties of soils from different land uses systems in Uttarakhand, India, *Journal of Chemical Engineering and Chemistry Research*, 3(11): 1114-1118.

Shishir Tandon

Affiliation: G. B. Pant University of Agriculture & Technology, Pantnagar 263 145 U. S. Nagar, (Uttarakhand), India.

Research and Professional Experience: 18 years of Research experience

Professional Appointments: Working as Scientist/Research Officer at G. B. Pant University of Agriculture & Technology, Pantnagar

Honours/Awards: Dr Shishir Tandon work is well recognized and was awarded twice as the Best Centre Award in 2006 & 2015 and Best

Annual Report Award in 2010. He has been awarded for Best Poster Award for his contribution in two different categories in National Symposium 2011. He has been awarded as one of the best participant by ICAR-IARI in 2017-18.

Publications from Last 23 Years:

Joshi, P. K., Joshi N., Tiwari, G. and Tandon, S. (2015). *Journal of Indian Chemical Society*, 92: 1603-1615.

Kumar, S, Tandon S. and Sand, N. K. (2014). *Bulletin of Environmental Contamination and Toxicology*, 92(2): 165-168.

Sharma, N., Sharma, N, Pant, R. and Tandon S. (2017). *Indian Journal of Plant Protection,* 45(4): 1-5.

Sharma. N. Pant R. and Tandon, S. (2015). *Oryza-An International Journal of Rice*, 52(2): 144-147.

Singh, A, Tandon, S. and Sand, N. K. (2014). *Journal of Chromatography and Separation Techniques*, 6(1): 257-259.

Singh, A.Tandon, S. and Sand, N. K. (2017). Asian Journal of Chemistry. 29(2): 271-273.

Singh, N. and Tandon, S. (2015). *International Journal of Environmental Science and Technology*, 12(8): 2475-2484.

Tandon, S. (2014). *Pest Management Science*, 70:1706-1710.

Tandon, S. (2015). *Plant Soil and Environment*. 61(11): 496-500.

Tandon, S. (2016). *Bulletin of Environmental Contamination and Toxicology*, 96(5): 694-698.

Tandon, S. (2017). *Indian Journal of Plant Protection,* 45(4): 1-5.

Tandon, S. (2017). *Journal of Food Agriculture & Environment* (Finland) 15(3&4): 98-101.

Tandon, S. (2018). *Indian Journal of Weed Science*.,50(2): 189-191.

Tandon, S. (2018). *International Journal of Environmental Analytical Chemistry*. 97(14-15): 1352-1361.

Tandon, S. (2019). Herbicide residues in rice–wheat cropping system in Uttarakhand. In *Herbicide Residue Research in India*. 1st edition, S.

Sondhia, P.P. Choudhury, and A.R. Sharma (Eds). Springer Nature Singapore Pte Ltd, Springer Verlag, Singapore, 253-260.

Tandon, S. (2019). *Journal of Food Protection*, 82(11):1959-1964.

Tandon, S. and Dubey, A. (2015). *Communications in Soil Science and Plant Analysis*. 46(7): 845-858.

Tandon, S. and Mittal A. K. (2017). *Journal of Chemistry*, (6): 1-3.

Tandon, S. and Mittal, A. K. (2018). *Biochemical Systematics and Ecology*, 81: 70-73.

Tandon, S. and Pant, R. (2019). *Environmental Technology*, 40(1): 86-93

Tandon, S. and Sand, N. K. (2015). *Pestology*, 39(8):32-34.

Tandon, S. and Sand, N. K. (2016). *International Journal of Medicinal Plants and Natural Products*, 2(1): 27-30.

Tandon, S. and Singh A. (2015). *Chemistry and Ecology*, 31(3): 273-284.

Tandon, S. and Singh, N. (2015). *Journal of Liquid Chromatography & Related Technologies*, 38(9): 993-996.

Tandon, S., Garg, G., Agarwal, R., and Sand, N. K. (2016). *Journal of Natural Products and Biomedical Research*. 2(1): 1-3.

Tandon, S., Kumar, S., and Sand, N. K. (2015). *International Journal of Analytical Chemistry,* 1-5.

Tandon, S., Mehra, P. and Sand, N. K. (2016). *Indian Journal of Weed Science*, 48(2): 230-232.

Tandon, S., Shah, S. K., Dubey, A and Verma, A. K. (2018). *Indian Farmer Digest*. 51(8): 25-28.

Trivedi, N., Tandon, and Dubey, A. (2017). *African Journal of Biotechnology*. 16(27): 1507-1512.

Vasmatkar, P., Dubey, A., Tyagi, B., Baral, P., Tandon, S. and Kadam, A. (2014). *Indo American Journal of Pharmaceutical Research*, 4(1): 304-311.

Verma, A. K., Dubey, A., and Tandon, S. (2015). *Indian Farmer Digest*, 48(2):37-38.

In: *Origanum*
Editor: Roger Ingram

ISBN: 978-1-53619-236-0

Chapter 2

ORIGANUM MEDIATED GREEN SYNTHESIS OF NANOPARTICLES

Isha Gupta*[1,2]*, Sonia Gandhi*[1,*]*, PhD and Sameer Sapra*[2]*, PhD

[1]Metabolomics Research Facility, Division of Behavioural Neuroscience, Institute of Nuclear Medicine and Allied Sciences, Defence Research and Development Organization, Delhi, India
[2]Department of Chemistry, Indian Institute of Technology, Delhi, India

ABSTRACT

Nanotechnology is gaining pace in the field of Science with each passing day due to various nanomaterials such as nanoparticles, nanotubes, nanorods, and hydrogels. These materials showcase versatile properties compared to their bulk counterparts, such as optical, anti-microbial, and magnetic. Different synthesis routes have been investigated to synthesise nanoparticles such as sol-gel (most common), spinning, pyrolysis, chemical vapour deposition, green synthesis,

[*]Corresponding Author's E-mail: sonia@inmas.drdo.in.

lithography, milling, sputtering, laser ablation, and thermal decomposition. While in all these synthesis routes, green synthesis is the eco-friendly, non-toxic, cost-effective, safe, and feasible route. Plant-mediated green synthesis is the simplest and most convenient method because of the use of metal salts and part of the plant as the precursors in the reaction. The extract prepared from *Origanum* species is used as a bio reductant because of phenolic compounds, act as a reducing agent in the synthesis process. The phytochemicals are also responsible for the stabilisation of the nanoparticles.In thischapter, *Origanum*'s role in synthesising the metal nanoparticles such as titanium dioxide,palladium, silver, gold, palladium nanoparticles supported on magnetic graphene oxide has been discussed in detail. Hence, researchers are using *Origanum* as a precursor in plant-mediated synthesis.

Keywords: *Origanum*, nanoparticles, green, phytochemicals, inorganic

INTRODUCTION

The Greek term '*nano*' means dwarf and represents the 10^{-9} m (one-billionth metre). One nanometer constitutes around three atoms. According to the International Organization for Standardization, the nanoparticles have a minimum one dimension in the range of 1-100 nm. In contrast, the definition given by the British Standards Institution states that all dimensions should be in the range of nanometre (Das and Chatterjee 2019). There are already many naturally occurring substances in our surroundings, such as deoxyribonucleic acid (DNA), parovirus, and proteins (Dolez 2015). In ancient times, the artisans and the architects were using nanotechnology without knowing it. Their work's brilliant example is the Lycurgus Cup having a stained glass with gold and silver nanoparticles, cementite nanowires inside carbon nanotubes (CNT). They all were exceptional in their work, but they did not know the Science behind it (I. Khan, Saeed, and Khan 2019). Accidently in 1857, Michael Faraday synthesised gold colloidal particles while mounting thin gold leaf on slides (Jeevanandam et al. 2018). Lately, in 1910, without knowing the mechanism, carbon black was used to reinforce tires because carbon black binds with the rubber to increase the hardness, strength, and other properties. "Nanotechnology" term was first time used by famous Richard Phillips Feynman in his talk of 1959, and there he said that "*There is plenty of room at the bottom*"(I. Khan, Saeed, and Khan 2019). Nanotechnology is the branch of Science that deals with developing and applying nanoparticle-based materials in different fields. The properties of nanoparticles are different from their bulk counterparts due to large surface area to volume ratio and the quantum confinement effects (Hong 2019). Hence, presently researchers are more invested in developing application-based technology having the size in the range of an atom. They are targeting fields such as medicine, energy, environment, and space. The particles have different morphologies such as spheres, cylinders and rods, crystallinity, and agglomeration, which can be synthesised from different chemical compounds such as metal and its oxides, polymers, carbon, and biomolecules. Nanoparticles are categorised based on composition as well as dimensions. Based on composition, nanoparticles are of three types: organic, inorganic, and carbon-based particles, while based on dimensions, nanoparticles are of three types: one-dimensional (1D), two-dimensional (2D), and three-dimensional (3D) (Ealias and Saravanakumar 2017; I. Khan, Saeed, and Khan 2019). The different nanoparticles are films (1D),carbon nanotubes (2D), dendrimers and quantum dots (3D) (Chokkareddy and Redhi 2018). Fullerenes and carbon nanotubes are two major divisions under carbon-based nanoparticles which are known for their versatile properties such as high electrical conductivity, and high electron affinity. Inorganic nanoparticles constitute metal and metal oxide. These nanoparticles show localised surface plasmon resonance, responsible for their exceptional optoelectronic properties. This property is due to the surface plasmon resonance, which is the excitation of free electrons present in the conduction band to the valence band. Organic nanoparticles are dendrimers, liposomes, micelles and ferritin. These nanoparticles are used in biomedical fields due to their non-toxic and biodegradable nature (Ealias and Saravanakumar 2017; I. Khan, Saeed, and Khan 2019).

Nanoparticles show various physiochemical properties such as optical, mechanical, large surface area and chemical, which is different from their bulk counterparts. The properties such as wettability, optical nature, and thermal conductivity are tunable according to the need and application (Jeevanandam et al. 2018). The particles have different morphologies, crystallinity and agglomeration. The optical properties of the inorganic nanoparticles are dependent on the size of particles. These particles show characteristic absorbance in the Ultraviolet-visible (UV-Vis) spectroscopy analysis. Most of the quantum dots show optical properties because the diameter is less than 10 nm. Silver and gold nanoparticles are yellow and rusty colours respectively. Magnetic properties of nanoparticles can be exploited in many fields such as biomedicine, magnetic resonance imaging (MRI), water decontamination. This property is noticeable for those particles with the size around 10-20 nm, particularly dependent on the synthesis route adopted and nanoparticles' electronic distribution(I. Khan, Saeed, and Khan 2019). Properties are dependent on their route, morphologies and the chemical agents or surfactants surrounding the nanoparticles. Surfactants easily modify interfacial properties. Much other molecular bonding comes into the picture after coming in the range of nanometers such as Van der Waals forces, electrostatic or covalent interactions which govern the particle aggregation. Therefore, particle-particle and particle-surrounding solvent interactions are important (Hong 2019).

Similarly, mechanical properties are applied in different fields, such as nanofabrication. This property includes hardness, elastic modulus, aggregation, surface coating, strain and stress. The inorganic nanoparticles are known for their thermal conductivity. Nanofluids are being developed for high heat transfer ability due to their large surface area and formed by dispersing the nanoparticles in different fluids such as glycol, oil or water (Khan, Saeed, and Khan 2019). Other physicochemical properties are surface-enhanced Raman scattering, high thermal and electrical conductivity (Parveen, Banse, and Ledwani 2016).

The nanoparticles are being exploited and developed for application in various fields such as biomedical engineering, drug delivery system, waste-water treatment, packaging, semiconductors, paint, cosmetics, biosensors, automobiles, and nano-fabrics, aerospace, energy, defence, textile, food, catalysis which is shown in Figure 1. The reason behind the application is the properties as discussed in the above paragraph. The conventional drug delivery system faces many problems, such as poor biodistribution and side effects. In contrast, the nano drugs have many advantages such as large surface area, better solubility, bioavailability, reduced dose requirement, decreased drug resistance, and improved drug dissolution rate. Amphotericin B nanotubes have been developed from carbon nanotubes which are more effective than the standalone drug. Gold nano shells can be used for diagnosis of immunoglobulins when couple with antibodies (Bhatia 2016). In electronics, the nanoparticles are coated without even melting. (Thakkar, Mhatre, and Parikh 2010) Nanoparticles can successfully convert the waste produced from food and agriculture (Parveen, Banse, and Ledwani 2016).The nanoparticles are effective in solar radiation when used in photovoltaic cells. Hence, nanotechnology has revolutionised all the fields due to its tunable properties (Parveen, Banse, and Ledwani 2016). Magnetic nanoparticles can be used in treating hypothermia based tumour. Nanoparticles remove toxic metals for groundwater treatment (A. K. Singh 2016).

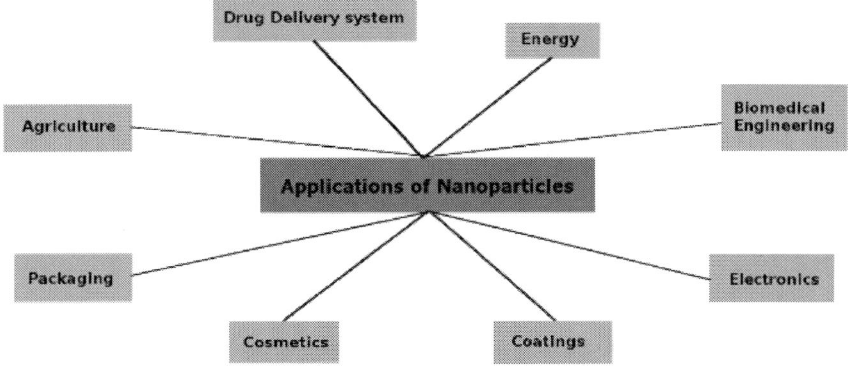

Figure 1. Applications of Nanoparticles.

This chapter briefly discusses the synthesis routes employed for nanoparticle development, emphasising the preferability of the green route over other routes, then the detailed discussion on different green route approaches. This chapter focuses on the increasing use of *Origanum* due to its geographical variation and rich composition. It discusses the recent development related to the *Origanum* mediated synthesis of inorganic nanoparticles such as palladium and silver and the constitution of *Origanum* herbs. The probable mechanism involved in the synthesis has also been discussed. The application of these synthesised nanoparticles in different fields of technology is leading to the development of the eco-friendlier approach in Science and Technology.

Synthesis Routes of Nanoparticles

Many synthesis routes have been employed for nanoparticles. The synthesis routes are broadly categorised into Top-down and Bottom-up. These two routes for synthesis include processes which are illustrated in Figure 2 (Ealias and Saravanakumar 2017). These are further divided into three sub-divisions: Physical, chemical and biological.

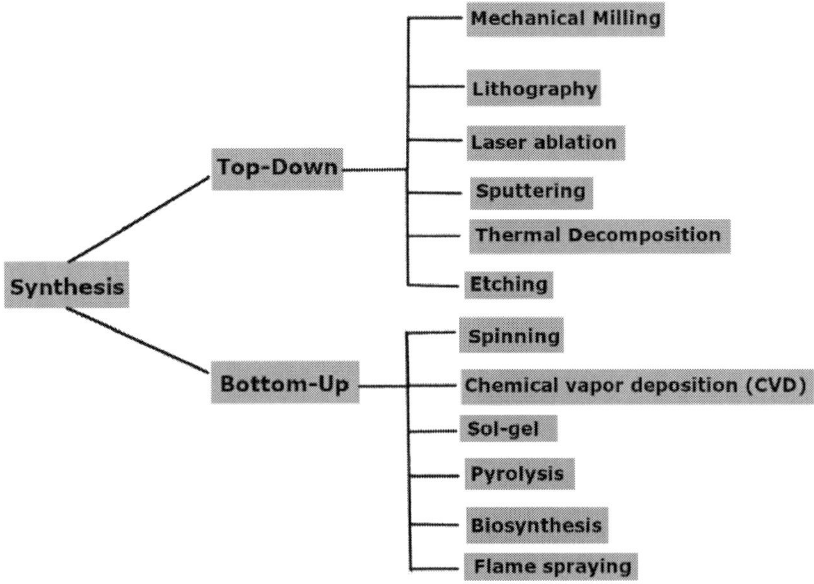

Figure 2. Synthesis of nanoparticles.

Bottom-up Process

The bottom-up process is also known as a constructive process because atoms build up to give a cluster of nanoparticles. The bottom-up process includes different processes such as sol-gel (most common), spinning, pyrolysis, chemical vapour deposition, and biosynthesis. The atoms are assembled systematically due to different interactions such as π-π interactions, hydrogen bonding, and Van der Waal forces. This process is based on the molecular recognition principle. This process is a low cost, easy and fast. Advantages of this process are fewer defects, homogeneous composition and production of uniformly shaped nanoparticles. This process needs more research and development to increase its application.

Top-Down Process

The top-down process is also known as a destructive process because the bulk is reduced to the atomic scale. The top-down process includes lithography, milling, sputtering, laser ablation and thermal decomposition. This process is quite simpler than the bottom up. Industries are dominantly using the top-down process to fabricate nanomaterials, such as for semiconductors and sensors. The nanomaterial is the miniature of bulk counterpart at the atomic level. Here, structure defects affect the physical properties of the nanoparticles. This process requires expensive and large installations and the formation of non-uniform shaped nanoparticles. Hence, the top-down process is not preferable due to the limitations mentioned above (Iqbal 2012).

Researchers mostly face problem in controlling the morphology, distribution, quality and quantity of nanoparticles. The morphology of nanoparticles is governed by factors involved in the synthesis such as temperature, pressure, pH, and environment. For example, high working temperatures are needed for the physical synthesis compared to ambient temperature for green synthesis.

GREEN SYNTHESIS OF NANOPARTICLES

The green synthesis is the safest and most convenient method than the other synthesis routes, explained in Figure 3 (Makarov et al. 2014; Kuppusamy et al. 2016; Parveen, Banse, and Ledwani 2016). This synthesis is in full accordance with the principles of green chemistry, helps to produce nanoparticles of desired shape and structure. This process comes under the category of the bottom-up process. Different types of precursors can be used, such as microorganisms, plants and templates such as proteins, lipids and carbohydrates. Microorganisms include fungi, bacteria, algae, and virus, while plants include parts such as leaves, tuber, stems, roots, bark, seeds, peel, flower, and callus (Ealias and Saravanakumar 2017; I. Khan, Saeed, and Khan 2019). Different energy resources are used to carry out the reaction, such as solar energy, microwave-assistance, ultrasound-mediation, hydrothermal. The synthesis can be done intracellularly and extracellularly from the microorganisms (G. Ingale 2013).

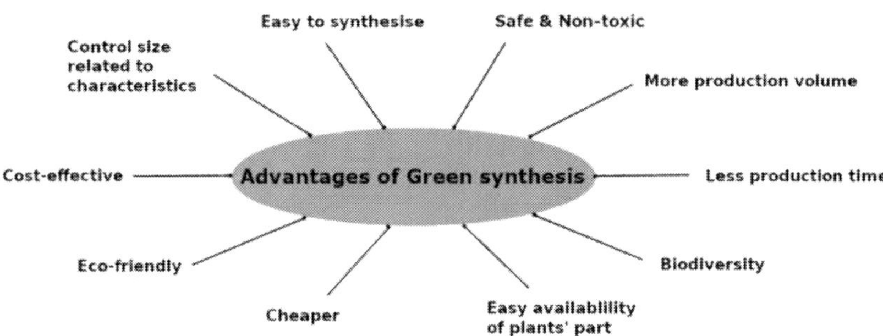

Figure 3. Advantages of green synthesis.

This synthesis is preferable for synthesising biocompatible inorganic nanoparticles. Metal nanoparticles are synthesised by bottom-up processes including sol-gel, chemical vapour deposition (CVD), biosynthesis while top-down including milling, lithography and sputtering. Similarly, metal oxide nanoparticles are made by bottom-up processes such as sol-gel, pyrolysis and biosynthesis in contrast to top-down such as milling, laser ablation and thermal decomposition (Ealias and Saravanakumar 2017). The chemical method has been widely used, involving solvent, reducing agents such as hydrazine, citrate, boranes, tetraoctylammonium bromide, polols, sodium borohydride, methoxy polyethylene glycol and stabilising agents such as polyvinyl alcohol. The chemicals, as mentioned above, are toxic to nature and living organisms. Therefore this approach cannot be used if we plan to use nanotechnology for humankind (Thakkar, Mhatre, and Parikh 2010).

Limitations of green syntheses are the production of a specific shape, size, monodisperse nanoparticles, and their toxicity on accumulating inside the organism's body. Standalone silver and copper nanoparticles can be toxic at a particular concentration (Parveen, Banse, and Ledwani 2016).

Microorganisms Based Synthesis

This process is also a good alternative for chemical synthesis due to its easy maintenance, and production. It has many advantages, such as fast, genetic diversity, availability, no harsh chemicals and influence on nanoparticles' properties (Das and Chatterjee 2019). Microorganisms have reductase enzymes that can reduce metal salts by accumulation and detoxification to produce specific size nanoparticles. In microorganism mediated synthesis, two methods are performed: Extracellular and intracellular. Extracellular synthesis is preferred over intracellular due to no downstream processing of synthesised nanoparticles is required.In downstream nanoparticles' processing, purification of nanoparticles is important and done by sonication, centrifugation and washing. The reducing agents such as peptides, proteins, genes, cofactors, and enzymes result in extra nanoparticles' extra stability(P. Singh et al. 2016).

1) Bacteria mediated synthesis: The reason behind using bacteria for synthesis is the ability to manipulate. Bacteria such as *Pseudomonas deceptionensis*, *Bhargavaea indica*, *Bacillus*

methylotrphicus, *Weissella oryzae*, *Bacillus*, *Escherichia*, *Streptomyces anulatus* and *Pyrobaculum* have been used in this synthesis (P. Singh et al. 2016). Slawson et al. synthesised silver nanoparticles of size ranging from 35 to 46 nm within the bacterial strain *Pseudomonas stutzeri* AG259 from silver mines (Slawson et al. 1994). The mechanism behind the synthesis is still not discovered. Sweeney et al. incubated the cadmium chloride and sodium sulphide with *Escherichia coli* to obtain cadmium sulphide nanocrystals with the improved property (Sweeney et al. 2004).

2) Yeast mediated synthesis: Few studies have been conducted with the yeast. Kowshik et al. synthesised silver nanoparticles with size ranging from 2 to 5 nm from yeast MKY3 species. This process was performed extracellularly followed by downstream processing (Kowshik et al. 2003). Dameron et al. reported cadmium nanoparticles from *Schizosaccharomyce pomble* and *Candida glabrata* (Dameron et al. 1989).
3) Fungi mediated synthesis: There are many advantages of using fungi: metal bioaccumulation, tolerance, secrete extracellular enzymes, cheaper, easy to handle and synthesise. It has only one limitation that is over-expression of any enzyme due to manipulation in genes than the prokaryotes. Mukherjee et al. reported intracellular silver nanoparticles of size 25 ± 12 nm from the Verticilium fungus. The SEM images supported that the silver nanoparticles were formed by the enzymes present in the cell wall (Mukherjee et al. 2001). Bhainsa et al. were able to synthesise stable silver nanoparticles extracellularly from *Aspergillus fumigatus* fungus, a pathogen for humans and plants.The reaction completed within minutes (Bhainsa and D'Souza 2006). The problem faced with intracellular synthesis is the processing of nanoparticles from fungus while extracellular synthesis is the microbe's disposal and handling(Thakkar, Mhatre, and Parikh 2010).
4) Virus mediated synthesis: Reports related to mineralisation of materials in cowpea mosaic and cowpea chlorotic mottle virus has

been reported.Studies of this type prove that nanoparticles can be nucleated and assembled on the virus's template(Douglas and Young 1998; Douglas et al. 2002).

Plant-Based Synthesis

Plant-based synthesis is the simplest, reliable and eco-friendly, which has led to the development of a new branch "Phytonanotechnology." Tremendous diversity in plants has led to the application in agriculture, pharmaceutical, and industry (Kavitha et al. 2013). As per some findings, metal bioaccumulation has been reported. These metals are accumulated in the nano range, such as when *Brassica juncea* (mustard greens) and *Medicago sativa* (alfalfa) were grown on silver nitrate, the silver nanoparticles (size of 50 nm) was found to have occupied 13.6% of the plants' body. Similarly, gold nanoparticles (size of 4 nm) were found in *Medigo sativa* (alfalfa) (Makarov et al. 2014).

Many papers have been published on using plant extracts for the synthesis of inorganic nanoparticles. This process involves metal salt and the plant's part as a precursor. This route is used in synthesising different inorganic nanoparticles such as silver, copper, gold, palladium, cobalt, platinum, magnetite, titanium dioxide and zinc oxide. Researchers used extracts from different plants and herbs such as *Cynara scolymus*, *Berberis* vulgaris, *Emblica Salvia soinosa*, *Tradescantia pallide*, *Paulownia tomentosa*, *Coriandrum sativum*, *Moringa oleifera*, *Diospyros montana*, *Annona reticulate*, *Enicostemma axillare*, Turmeric, *Melissa officinalis*, *Citrus sinensis*, *Origanum majorana*, Apple, *Clitoria ternate*s, *Solanum nigrum*, *Aloe vera, Pimenta dioica*, *Murraya koenigii*, *Alternanthera sessilis*, *Morinda citrifolia*, *Boswellia serrata*, *Ocimum sanctum*, *Triticum aestivum*, *Gliricidia sepium*, *Medigo sativa*, and *Azadirachta indica*.

This route is an *in vitro* approach using different parts of plants such as tuber, roots, callus, stems, fruits, and peels, used as a precursor in the reaction. The extract contains primary metabolites such as vitamins, amino acids, proteins, organic acids and secondary metabolites such as terpenoids, flavonoids, phenolic compounds, alkaloids, and reducing sugar. These metabolites are responsible for the redox reaction to synthesise and stabilise the synthesised nanoparticles. The different plants have different compositions and concentrations of these metabolites due to which the morphology of nanoparticles changes. The nanoparticles' morphology can be easily controlled by altering reaction conditions such as temperature, pH, concentration, and mixing ratio by volume.

MECHANISM INVOLVED IN PLANT-MEDIATED SYNTHESIS

Researchers have tried to explain the synthesis of nanoparticles mediated by green routes. In this route, the plant extract is added to the metal salt precursors under different conditions. These conditions include the concentration of the extract, phyto-compounds, and metal salt, pH and temperature. The phyto-compounds affect the shape and size of nanoparticles. The flavonoids reduce the metal salt by giving hydrogen atom from the tautomerism of enol to keto form. Amino compounds have been found to be associated with the metal ions in order to reduce it to metal (0). The functional groups such as –C-O-, –C=O, -COC-, and –C=C- are responsible for stabilising the nanoparticles (Singh, 2018).

The synthesis consists of three steps: Activation, growth, and termination. First, the activation step involves reducing the metal ions of metal salts to its respective metal atom, followed by nucleation of the atoms. In this, metal ions in metal salts are reduced to metal (zero-oxidation state) by plant metabolites. Second, the growth phase involves the coalescence of the particle with the neighbouring particles increasing thermodynamic stability. Third, the termination step involves the development of the shape of nanoparticles. These particles can aggregate to form different shapes, such as nanorods, nanotubes, nanohexadrons and nanoprisms or irregular shapes. Hence, in the last step, the nanoparticles attain the most stable conformation due to the metabolites' stabilisation effect present in plant extracts, explained in Figure 4. The nanoparticles tend to achieve the most stable shape to minimise its Gibbs free energy

(Makarov et al. 2014; Kuppusamy et al. 2016; Chokkareddy and Redhi 2018).

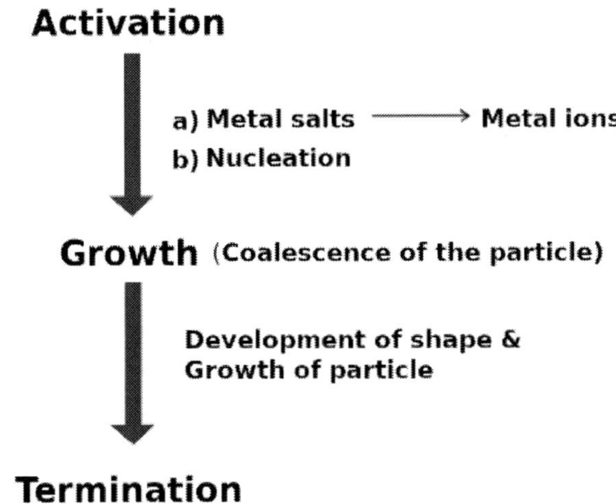

Figure 4. Mechanism of synthesis.

Characterisation of Nanoparticles

The characterisation of synthesised nanoparticles is an important step to determine the properties of the particles. Researchers use different techniques to identify the properties such as Scanning electron microscopy (SEM), Transmission electron microscopy (TEM), Electron dispersive X-ray spectroscopy (EDX), X-ray diffraction (XRD), Brunauer–Emmett–Teller (BET), Fourier transform-Infrared (FTIR) spectroscopy and Thermogravimetric analysis.

Morphology of Nanoparticles

This is a basic and important parameter which directly regulate the size, shape and distribution of particles. It has been reported that different synthesis routes result in different morphologies of the same nanoparticles.

The morphology is studied with the electron microscopic techniques such as Scanning electron microscopy (SEM), Transmission electron microscopy (TEM) and polarised optical microscopy (POM). SEM provides information regarding the morphology as well as the surroundings of the nanoparticle. The limitation is lack of distribution information. Here, secondary electrons are used for this purpose. TEM provides information related to size and diffraction. This microscopy involves time-consuming sample preparation process. Energy Dispersive X-ray spectroscopy and Electron-energy loss spectroscopy are hyphenated with high-resolution TEM to obtain more data. Scanning Mobility Particle Sizer is used for gaseous nanoparticles, known for fast and accurate results. Atomic force microscopy (AFM) is also known as scanning force microscopy because it helps obtain the images of the specimen surfaces. Samples are scanned in two modes: contact and non-contact. The main advantage of AFM is that it can image the non-conducting samples without any preparation. Hence, this technique can be used for microstructures, polymers and biological samples (Chokkareddy and Redhi 2018; Hong 2019; Nethi et al. 2019).

Structure

In nanoparticles' structure, the composition, nature of bonding, and presence of undesired element or molecule are analysed. Different techniques are used for the determination of structure such as X-ray diffraction (XRD), Energy Dispersive X-ray spectroscopy (EDX), X-ray photoelectron spectroscopy (XPS), Fourier transform-Infrared (FTIR) spectroscopy, Raman spectroscopy, Brunauer–Emmett–Teller (BET), Zeta size analyser. XRD helps in studying crystallography, i.e.,the arrangement of atoms and molecules in a crystal. In this way, crystallinity and the phase of nanoparticles are studied. Particle size can be analysed with the Debye Scherrer formula. Surface area studies are done with the help of BET analysis, Nuclear Magnetic Resonance (NMR) spectroscopy, modified Scanning Mobility Particle Sizer spectrometer (SMPS). EDX spectroscopy reinforces the data obtained from the electron microscopy such as SEM.

This spectroscopy gives insight into the elemental composition of the particles. XPS is a well-known surface-sensitive technique used to demonstrate the elements' bonding and elemental ratio and the depth profile. FT-IR and Raman spectroscopy arethe most feasible techniques among all. Zeta size analyser or Dynamic light scattering (DLS) analyser studies the size variation, but the main limitation behind this technique is that it gives the hydrodynamic diameter. In this technique, the size calculation is based on the Brownian motion of nanoparticles in the dispersion medium when illuminated by monochromatic light. So, change in the wavelength relates to the size of nanoparticles. Other parameters of nanoparticles, such as diffusion constant and zeta potential, are also calculated. Determination of zeta potential helps calculate the surface charge, which ultimately tells the stability of a particle. This technique is the fastest and accurate.The data obtained from DLS support the data obtained from the SEM(Bhatia 2016; Hong 2019).

Optical Characterisation

This characterisation covers absorption, reflectance, luminescence and phosphorescence for photocatalytic application. The bandgap of any materials is crucial to understand the conductance and photo-activity of the material. Instruments such as UV-Vis spectroscopy, null ellipsometer and photoluminescence (PL) are employed to study this nanoparticles' characteristic. The UV-Vis spectroscopy can measure the transmittance, absorbance and reflectance, which ultimately help to know the bandgap of the synthesised nanoparticles.In contrast, PL helps determine the absorption and emission capacity of any material (Hong 2019).

ORIGANUM

Plant-mediated synthesis is being exploited in various fields such as food and anti-microbial agent because the phytochemicals present in plants part act as reducing and capping agent. Therefore many studies have been published in the last two decades on the above mentioned traditional herbs. Few studies have been conducted on the extract formed from *Origanum.*

Origanum, which is popularly known as Oregano, belongs to the Lamiaceae family, found in some of the Mediterranean and western and southwestern Eurasia regions. This plant is dense, tubular, erect, five-lobed calyces, ascending and highly aromatic. It is a culinary herb due to its usage as a flavouring agent in food and alcohol. This herb is one of the medicinal and allopathic plants which of almost 42 species. As per Letswaart, *Origanum* is classified into ten divisions based on the morphology: Amaracus Bentham, Anatolicon Bentham, Bervifilamentum letswaart, Longitubus letswaart, Chilocalyx letswaart, Majorana Bentham, Campanulaticalyx letswaart, Elongatispica letswaart, *Origanum*, and Prolaticorolla letswaart. The most abundant and monospecific division is *Origanum*, found in other regions such as North Africa and Eurasia as well. It is further divided into six subspecies based on indumentum, sessile glands, and bracts: hirtum, *Vulgare*, glandulosum, gracile, viridulum, and virens. In India, O. *Vulgare* (L.) is known as Himalayan marjoram or jungali tulsi (Vokou, Kokkini, and Bessiere 1993; Stella Kokkini, Karousou, and Vokou 1994; S. Kokkini, Karousou, and Hanlidou 2003; Chishti, Kaloo, and Sultan 2013).

Oregano's dried leaves and essential oil are being used in the therapeutic field for its tonic, diaphoretic, stimulant, anti-parasitic, antioxidant, anti-thrombin, anti-inflammatory, antispasmodic,carminative, and antiseptic properties. Therefore, it is part of traditional medicine systems for treating disorders such as urinary tract, dental caries, digestive, and respiratory. It is also used in cosmetic and agricultural fields (Chishti, Kaloo, and Sultan 2013; Mohamed 2014; Benedec et al. 2018). Different studies have been performed with this genus.

PHYTOCHEMICALS IN *ORIGANUM*

As per the World Health Organisation report, the population of more than 80% of the world is dependent on traditional herbal medicine. Therefore, phyto-compounds present in the plant is gaining the focus of researchers. They are also extracting these metabolites with the help of tissue culture. The secondary metabolites are the Phyto-compounds, as these small molecules are intermediates and products of metabolism. Molecules perform different roles such as signalling, inhibitory and stimulatory effects, catalytic effects, and defence against the pathogens. The traditional herbs are rich in these molecules. The secondary metabolites are broadly classified into four divisions: Alkaloids, phenols, flavonoids, and steroids (Tiwari and Rana 2015; Jain, Khatana, and Vijayvergia 2019).

The phytochemicals present in *Origanum* are diverse due to the environmental and genetic conditions. Sarikurkcu et al. studied the composition of *O. vulgare* subsp. *vulgare* and *O. vulgare* subsp. *hirtum*. As per their findings, *O. vulgare* subsp. *vulgare* contains a significant amount of thymol followed by carvacrol and p-cymene. The other components are borneol, α-pinene, α-thujene, camphene, α-terpinene, γ-terpinene, 1-octen-3-ol, β-caryophyllene, β-bisabolene, p-cymene-8-ol, and caryophyllene oxide. *O. vulgare* subsp. *hirtum* contains around 96-97% of linalool. It has trans-linalol oxide, cis-linalol oxide, hotrienol, β-caryophyllene, α-humulene, α-terpineol, borneol, elemol and carvacrol in minor quantities. From the above information, we can conclude that each subspecies contain different amount of metabolites (Sarikurkcu et al. 2015). Verma et al. identified 38 components in the essential oil with the help of Gas Chromatography (GC) and Gas chromatography – Mass spectrometry (GC-MS). They found major content of monoterpenes at the full flowering stage of *O. vulgare* (Verma, Padalia, and Chauhan 2012). M. Khan et al. analysed the stems and leaves of *O. vulgare* from Jordan and Saudi Arabia by GC-MS. This research helped to identify almost 153 metabolites. As per their results, carvacrol and thymol were major constituents in the Saudi and Jordan species, respectively (M. Khan et al. 2019).

The essential oil of *Origanum* contains around 71% of phenolic compounds. Carvacrol, a monoterpenoid phenol, present in high amount in O. *vulgare* subspecies. The *in vitro* and *in vivo* studies have shown that this compound has anti-cancer, anti-microbial, anti-fungal, anti-viral, antioxidant, and antispasmodic effects. The other monoterpenes such as sabinene linalyl acetate and cis-sabinene hydrate are found in the species *O. majorana* L. These metabolites have many applications such as common cold, anti-rheumatic and spasmolytic. Hence, *Origanum* has higher phenolic content than other herbal medicine (Mohamed 2014). Coccimiglio et al. identified the constituents present in the ethanolic extract of *O. vulgare* by GC-MS, which also proved thymol and carvacrol in significant quantities while p-cymene, 1-phytol, cresol, and octacosanol in minor quantities. This extract showed cytotoxic, antibacterial and antioxidant activities due to the presence of thymol and carvacrol (Coccimiglio et al. 2016).

ORIGANUM MEDIATED SYNTHESIS OF NANOPARTICLES

Many researchers have reported the synthesis of inorganic nanoparticles such as palladium, titanium dioxide, silver, and gold from *Origanum*'s extract due to the presence of the diverse secondary metabolites as discussed earlier which act as reducing as well as a capping agent in the synthesis process.

Sankar et al. used leaves extract of *O. vulgare* produce silver nanoparticles of 136 ± 10.09 nm. In the FTIR spectrum, bands at 3410 cm^{-1}, 2250 cm^{-1}, 1604 cm^{-1}, and 1048 cm^{-1} were of N-H stretching of C=N, bending of aromatic C=C,and stretching of C-O, respectively. These functional groups are involved in the reduction and stabilisation of the nanoparticles. They also studied dose-response against the cell line of human lung cancer A549 (Sankar et al. 2013).

Sankar et al. reported the efficacy in wound healing by *O. vulagre* mediated titanium dioxide nanoparticles in rats. The particle size of 341 nm was identified with the help of the Dynamic Light Scattering analyser.

In Wistar albino male rats, 94% wound closure was observed on treatment with titanium dioxide than 86% in control (Sankar et al. 2014).

Singh et al. performed the microwave-assisted synthesis of silver nanoparticles from the aqueous leaves to extract of *O. majorana*. In the FTIR spectrum, bands at 3368 and 1607 cm^{-1} were shifted due to the association of O-H and N-H with the silver nanoparticles. The SEM micrographs showed the formation of spherical nanoparticles having a size of less than 70 nm. The nanoparticles showed better antibacterial property against *E.coli* at the concentration of 25 $\mu g\ ml^{-1}$ than *B. subtilis* (Singh, Rawat, and Gupta 2016).

Rafi Shaik et al. synthesised thermally stable palladium nanoparticles from the extract of the aerial parts of *O. vulagre* L and palladium (II) chloride. The spherical shaped nanoparticles with the size ranging from 2-20 nm were produced from this approach. Using FTIR spectroscopy, phenolic compounds and glycosides were found responsible for reducing Pd(II) to Pd(0) because of the oxidation to carbonyl groups. These nanoparticles were used as a catalyst for the oxidation of alcohols (Rafi Shaik et al. 2017).

Seyedi et al. prepared palladium nanoparticles grafted on modified graphene oxide from aqueous extract of leaves of *O. vulgare*. The extract of O. *Vulgare* reduced the palladium ion (Pd2+) to palladium (Pd). The TEM studies showed the size of Pd nanoparticles ranging from 10-40 nm. The catalytic property was studied for reactions such as Suzuki-Miyaura cross-coupling and reducing dyes (methyl orange and methylene blue). GO/Fe3O4/Pd catalyst was preferable over other catalysts due to its recyclability (Seyedi, Saidi, and Sheibani 2018).

Zahran et al. used aqueous extract prepared from the fresh leaves of *O. majorana* for the preparation of silver nanoparticles. Silver nanoparticles have the size from 25 - 50 nm. In the FTIR spectrum, the bands at 3395, 2922, and 1639.97 cm^{-1} represent stretching modes of N-H and O-H bonds, C-H and C=O, respectively. N-H stretching band was found broader, which implies the coordination of N-H group with the silver nanoparticles. This research group also discussed the stability of the nanoparticles synthesised from *O. majorana*. They emphasised applying these nanoparticles as a coating material for the water treatment plant's pipes or reservoirs. These cyclic voltammetry studies showed nanoparticles at 0.39 mV (Zahran et al. 2018).

Benedec et al. prepared biocompatible gold nanoparticles from the aerial parts or leaves of *Origanum herba* or *Origanum folium* of *Origanum vulgare*. These nanoparticles possess enhanced plasmonic, anti-microbial, and antioxidant properties. Gold nanoparticles showed antimycotic and bactericidal action against *C. albicans* and *S. aureus*(Benedec et al. 2018).

Nezhad et al. synthesised cerium oxide from the leaf extract and cerium nitrate hexahydrate. These nanoparticles were imaged through TEM and found the formation of spherical shaped 20 nm particles. With the FTIR spectrum's help, it was deduced that phenolic groups and flavonoids were found with cerium oxide nanoparticles. The antioxidant activity of *Origanum* mediated synthesis cerium oxide was studied against the ABTS and DPPH. These nanoparticles showed anti-cancer properties against breast cancer (Nehzad, Es-haghi, and Tabrizi 2019).

The limited number of studies have been conducted with the *Origanum* herb. The extract prepared from this herb can synthesise other metallic nanoparticles such as titanium dioxide, cadmium, and zinc oxide, which are also known for widespread applications. In this way, greener methods will be used to develop and produce the nanoparticles, ultimately leading to eco-friendlier sustainable development.

Conclusion

The nanotechnology has revolutionised many fields such as health, environment, energy, packaging, cosmetics, drug delivery, automobiles and food due to its morphology and surroundings dependent properties. Mostly, nanoparticles are extensively synthesised by chemical and physical routes which are expensive, involve complex installations and toxic chemicals. Many researchers are constantly working for synthesising biocompatible, stable inorganic nanoparticles from plant-mediated green synthesis as a step towards green chemistry. This step will lead to the development of a new field of Science, "phytonanotechnology." This field involves the use of the rapid and single-step procedure for synthesis. Stable nanoparticles with controlled morphology are synthesised with the help of herbs' extract. The current review emphasises on the use of *Origanum* mediated green synthesis. Many researchers have done profiling of the secondary metabolites present in the different parts of the *Origanum,* which proved the presence of different phyto-compounds due to the environmental and genetic variations. This chapter has briefly discussed the studies conducted for nanoparticle preparations from the *Origanum* such as gold and silver nanoparticles. The only concerns hindering the large scale production are distribution profile, clearance and excretion of nanoparticles from the body need to be more studied. *Origanum* mediated green synthesis of nanoparticles can play an important role in the booming world of "phytonanotechnology."

REFERENCES

Benedec, Daniela, Ilioara Oniga, Flavia Cuibus, Bogdan Sevastre, Gabriela Stiufiuc, Mihaela Duma, Daniela Hanganu, Cristian Iacovita, Rares Stiufiuc, and Constantin Mihai Lucaciu. 2018. "*Origanum Vulgare* Mediated Green Synthesis of Biocompatible Gold Nanoparticles Simultaneously Possessing Plasmonic, Antioxidant and Anti-microbial Properties." *International Journal of Nanomedicine* Volume 13 (February): 1041–58. https://doi.org/10.2147/IJN.S149819.

Bhainsa, Kuber C., and S.F. D'Souza. 2006. "Extracellular Biosynthesis of Silver Nanoparticles Using the Fungus Aspergillus Fumigatus." *Colloids and Surfaces B: Biointerfaces* 47 (2): 160–64. https://doi.org/10.1016/j.colsurfb.2005.11.026.

Bhatia, Saurabh. 2016. Natural Polymer Drug Delivery Systems. *Natural Polymer Drug Delivery Systems*. Cham: Springer International Publishing. https://doi.org/10.1007/978-3-319-41129-3.

Chishti, Shayista, Zahoor A. Kaloo, and Phalestine Sultan. 2013. "Medicinal Importance of Genus *Origanum*: A Review."*Journal of*

Pharmacognosy and Phytotherapy 5 (10): 170–77. https://doi.org/10.5897/JPP2013.0285.

Chokkareddy, Rajasekhar, and Gan Redhi. 2018. "Green Synthesis of Metal Nanoparticles and Its Reaction Mechanisms." In *Green Metal Nanoparticles: Synthesis, Characterisation and Their Application; Kanchi, S., Ahmed, S. Eds*, 113–39. Hoboken, NJ, USA: John Wiley & Sons, Inc. https://doi.org/10.1002/9781119418900.ch4.

Coccimiglio, John, Misagh Alipour, Zi-Hua Jiang, Christine Gottardo, and Zacharias Suntres. 2016. "Antioxidant, Antibacterial, and Cytotoxic Activities of the Ethanolic *Origanum Vulgare* Extract and Its Major Constituents." *Oxidative Medicine and Cellular Longevity* 2016: 1–8. https://doi.org/10.1155/2016/1404505.

Dameron, C. T., R. N. Reese, R. K. Mehra, A. R. Kortan, P. J. Carroll, M. L. Steigerwald, L. E. Brus, and D. R. Winge. 1989. "Biosynthesis of Cadmium Sulphide Quantum Semiconductor Crystallites." *Nature* 338 (6216): 596–97. https://doi.org/10.1038/338596a0.

Das, Mousumi, and Saptarshi Chatterjee. 2019. "Green Synthesis of Metal/Metal Oxide Nanoparticles toward Biomedical Applications: Boon or Bane." In *Green Synthesis, Characterisation and Applications of Nanoparticles*, 265–301. Elsevier. https://doi.org/10.1016/B978-0-08-102579-6.00011-3.

Dolez, Patricia I. 2015. "Nanomaterials Definitions, Classifications, and Applications." In *Nanoengineering*, 3–40. Elsevier. https://doi.org/10.1016/B978-0-444-62747-6.00001-4.

Douglas, Trevor., Erica. Strable, Deborah. Willits, Abdelaziz. Aitouchen, Mathew. Libera, and Mark. Young. 2002. "Protein Engineering of a Viral Cage for Constrained Nanomaterials Synthesis." *Advanced Materials* 14 (6): 415–18. https://doi.org/10.1002/1521-4095(20020318)14:6<415::AID-ADMA415>3.0.CO;2-W.

Douglas, Trevor., and Mark. Young. 1998. "Host–Guest Encapsulation of Materials by Assembled Virus Protein Cages." *Nature* 393 (6681): 152–55. https://doi.org/10.1038/30211.

Ealias, Anu Mary, and M. P. Saravanakumar. 2017. "A Review on the Classification, Characterisation, Synthesis of Nanoparticles and Their

Application." *IOP Conference Series: Materials Science and Engineering* 263 (3): 032019. https://doi.org/10.1088/1757-899X/263/3/032019.

Ingale, G. Arun. 2013. "Biogenic Synthesis of Nanoparticles and Potential Applications: An Eco-Friendly Approach." *Journal of Nanomedicine & Nanotechnology* 04 (02): 7. https://doi.org/10.4172/2157-7439.1000165.

Hong, Nguyen Hoa. 2019. "Introduction to Nanomaterials: Basic Properties, Synthesis, and Characterization." In *Nano-Sized Multifunctional Materials*, 1–19. Elsevier. https://doi.org/10.1016/B978-0-12-813934-9.00001-3.

Jain, Chitra, Shivani Khatana, and Rekha Vijayvergia. 2019. "Bioactivity of Secondary Metabolites of Various Plants: A Review."*International Journal of Pharmaceutical Sciences and Research* 10 (2): 494–504. https://doi.org/10.13040/IJPSR.0975-8232.10(2).494-04.

Jeevanandam, Jaison, Ahmed Barhoum, Yen S. Chan, Alain Dufresne, and Michael K. Danquah. 2018. "Review on Nanoparticles and Nanostructured Materials: History, Sources, Toxicity and Regulations." *Beilstein Journal of Nanotechnology* 9 (1): 1050–74. https://doi.org/10.3762/bjnano.9.98.

Kavitha, K. S., Syed Baker, D. Rakshith, H. U. Kavitha, Yashwantha Rao H. C., B. P. Harini, and S. Satish. 2013. "Plants as Green Source towards Synthesis of Nanoparticles." *International Research Journal of Biological Sciences* 2 (6): 66–76.

Khan, Ibrahim, Khalid Saeed, and Idrees Khan. 2019. "Nanoparticles: Properties, Applications and Toxicities." *Arabian Journal of Chemistry* 12 (7): 908–31. https://doi.org/10.1016/j.arabjc.2017.05.011.

Khan, Merajuddin, Shams T. Khan, Mujeeb Khan, Ahmad A. Mousa, Adeem Mahmood, and Hamad Z. Alkhathlan. 2019. "Chemical Diversity in Leaf and Stem Essential Oils of *Origanum Vulgare* L. and Their Effects on Microbicidal Activities." *AMB Express* 9 (1): 176. https://doi.org/10.1186/s13568-019-0893-3.

Kokkini, S., R. Karousou, and E. Hanlidou. 2003. "HERBS | Herbs of the Labiatae." *Encyclopedia of Food Sciences and Nutrition*, 3082–90. https://doi.org/10.1016/b0-12-227055-x/00593-9.

Kokkini, Stella, Regina Karousou, and Despina Vokou. 1994. "Pattern of Geographic Variations of *Origanum Vulgare* Trichomes and Essential Oil Content in Greece." *Biochemical Systematics and Ecology* 22 (5): 517–28. https://doi.org/10.1016/0305-1978(94)90046-9.

Kowshik, Meenal, Shriwas Ashtaputre, Sharmin Kharrazi, W. Vogel, J. Urban, S. K. Kulkarni, and K. M. Paknikar. 2003. "Extracellular Synthesis of Silver Nanoparticles by a Silver-Tolerant Yeast Strain MKY3." *Nanotechnology* 14 (1): 95–100. https://doi.org/10.1088/0957-4484/14/1/321.

Kuppusamy, Palaniselvam, Mashitah M. Yusoff, Gaanty Pragas Maniam, and Natanamurugaraj Govindan. 2016. "Biosynthesis of Metallic Nanoparticles Using Plant Derivatives and Their New Avenues in Pharmacological Applications – An Updated Report." *Saudi Pharmaceutical Journal* 24 (4): 473–84. https://doi.org/10.1016/j.jsps.2014.11.013.

Makarov, V. V., A. J. Love, O. V. Sinitsyna, S. S. Makarova, I. V. Yaminsky, M. E. Taliansky, and N. O. Kalinina. 2014. "'Green' Nanotechnologies: Synthesis of Metal Nanoparticles Using Plants." *Acta Naturae* 6 (1): 35–44. https://doi.org/10.32607/20758251-2014-6-1-35-44.

Mohamed, Suhaila. 2014. "Herbs and Spices in Aging." In *Aging*, 99–107. Elsevier. https://doi.org/10.1016/B978-0-12-405933-7.00010-X.

Mukherjee, Priyabrata, Absar Ahmad, Deendayal Mandal, Satyajyoti Senapati, Sudhakar R. Sainkar, Mohammad I. Khan, Renu Parishcha, et al. 2001. "Fungus-Mediated Synthesis of Silver Nanoparticles and Their Immobilisation in the Mycelial Matrix: A Novel Biological Approach to Nanoparticle Synthesis." *Nano Letters* 1 (10): 515–19. https://doi.org/10.1021/nl0155274.

Nehzad, Saynaz Aseyd, Es-haghi, Ali, and Tabrizi, Masoud Homayouni. "Green synthesis of cerium oxide nanoparticle using *Origanum* majorana L. leaf extract, its characterisation and biological activities."

*Applied Organometallic Chemistry*34, no. 2 (2020): e5314. https://doi.org/10.1002/aoc.5314.

Nethi, Susheel Kumar, Sourav Das, Chitta Ranjan Patra, and Sudip Mukherjee. 2019. "Recent Advances in Inorganic Nanomaterials for Wound-Healing Applications." *Biomaterials Science* 7 (7): 2652–74. https://doi.org/10.1039/c9bm00423h.

Parveen, Khadeeja, Viktoria Banse, and Lalita Ledwani. 2016. "Green Synthesis of Nanoparticles: Their Advantages and Disadvantages." In *AIP Conference Proceedings*, 1724:020048. https://doi.org/10.1063/1.4945168.

Rafi Shaik, Mohammed, Zuhur Jameel Qandeel Ali, Mujeeb Khan, Mufsir Kuniyil, Mohamed E. Assal, Hamad Z. Alkhathlan, Abdulrahman Al-Warthan, Mohammed Rafiq H. Siddiqui, Merajuddin Khan, and Syed Farooq Adil. 2017. "Green Synthesis and Characterization of Palladium Nanoparticles Using *Origanum Vulgare* L. Extract and Their Catalytic Activity."*Molecules* 22 (1): 165. https://doi.org/10.3390/molecules22010165.

Sankar, Renu, Ravishankar Dhivya, Kanchi Subramanian Shivashangari, and Vilwanathan Ravikumar. 2014. "Wound Healing Activity of *Origanum Vulgare* Engineered Titanium Dioxide Nanoparticles in Wistar Albino Rats."*Journal of Materials Science: Materials in Medicine* 25 (7): 1701–8. https://doi.org/10.1007/s10856-014-5193-5.

Sankar, Renu, Arunachalam Karthik, Annamalai Prabu, Selvaraju Karthik, Kanchi Subramanian Shivashangari, and Vilwanathan Ravikumar. 2013. "*Origanum Vulgare* Mediated Biosynthesis of Silver Nanoparticles for Its Antibacterial and Anticancer Activity." *Colloids and Surfaces B: Biointerfaces* 108: 80–84. https://doi.org/10.1016/j.colsurfb.2013.02.033.

Sarikurkcu, Cengiz, Gokhan Zengin, Mustafa Oskay, Sengul Uysal, Ramazan Ceylan, and Abdurrahman Aktumsek. 2015. "Composition, Antioxidant, Antimicrobial and Enzyme Inhibition Activities of Two *Origanum Vulgare* Subspecies (Subsp. *Vulgare* and Subsp. Hirtum) Essential Oils." *Industrial Crops and Products* 70 (August): 178–84. https://doi.org/10.1016/j.indcrop.2015.03.030.

Seyedi, Neda, Kazem Saidi, and Hassan Sheibani. 2018. "Green Synthesis of Pd Nanoparticles Supported on Magnetic Graphene Oxide by *Origanum Vulgare* Leaf Plant Extract: Catalytic Activity in the Reduction of Organic Dyes and Suzuki–Miyaura Cross-Coupling Reaction." *Catalysis Letters* 148 (1): 277–88. https://doi.org/10.1007/s10562-017-2220-4.

Singh, A. K. 2016. "Structure, Synthesis, and Application of Nanoparticles." In *Singh AK. Engineered Nanoparticles-Structure, Properties and Mechanisms of Toxicity*, 19–76. Elsevier. https://doi.org/10.1016/B978-0-12-801406-6.00002-9.

Singh, Darshan, Deepti Rawat, and Isha Gupta. 2016. "Microwave-Assisted Synthesis of Silver Nanoparticles from *Origanum* Majorana and Citrus Sinensis Leaf and Their Antibacterial Activity: A Green Chemistry Approach." *Bioresources and Bioprocessing* 3 (1): 14. https://doi.org/10.1186/s40643-016-0090-z.

Singh, Priyanka, Yu-Jin Kim, Dabing Zhang, and Deok-Chun Yang. 2016. "Biological Synthesis of Nanoparticles from Plants and Microorganisms." *Trends in Biotechnology* 34 (7): 588–99. https://doi.org/10.1016/j.tibtech.2016.02.006.

Slawson, R. M., E. M. Lohmeier-Vogel, H. Lee, and J. T. Trevors. 1994. “Silver Resistance in Pseudomonas Stutzeri.” *Biometals* 7 (1): 30–40. https://doi.org/10.1007/BF00205191.

Sweeney, Rozamond, Chaunbin Mao, Xiaoxia Gao, Justin L. Brut, Angela M. Belcher, George Georgiou, and Brent L. Iverson. 2004. "Bacterial Biosynthesis of Cadmium Sulfide Nanocrystals."*Chemistry & Biology* 11 (11): 1553–59. https://doi.org/10.1016/j.chembiol .2004.08.022.

Thakkar, Kaushik N., Snehit S. Mhatre, and Rasesh Y. Parikh. 2010. "Biological Synthesis of Metallic Nanoparticles." *Nanomedicine : Nanotechnology, Biology, and Medicine* 6 (2): 257–62. https://doi.org/10.1016/j.nano.2009.07.002.

Tiwari, Ruby, and C.S. Rana. 2015. "Plant Secondary Metabolites." In *International Journal of Engineering Research and General Science*, 661–70. New Delhi: Springer India. https://doi.org/10.1007/978-81-322-2401-3_11.

Verma, Ram Swaroop, Rajendra Chandra Padalia, and Amit Chauhan. 2012. "Volatile Constituents of *Origanum Vulgare* L., Thymol Chemotype: Variability in North India during Plant Ontogeny." *Natural Product Research* 26 (14): 1358–62. https://doi.org/10.1080/14786419.2011.602017.

Vokou, Despina, Stella Kokkini, and Jean Marie Bessiere. 1993. "Geographic Variation of Greek Oregano (*Origanum Vulgare* Ssp. Hirtum) Essential Oils." *Biochemical Systematics and Ecology* 21 (2): 287–95. https://doi.org/10.1016/0305-1978(93)90047-U.

Zahran, Moustafa, Maged El-Kemary, Shaden Khalifa, and Hesham El-Seedi. 2018. "Spectral Studies of Silver Nanoparticles Biosynthesised by *Origanum* Majorana." *Green Processing and Synthesis* 7 (2): 100–105. https://doi.org/10.1515/gps-2016-0183.

In: *Origanum*
Editor: Roger Ingram
ISBN: 978-1-53619-236-0

Chapter 3

THE ROLE OF *ORIGANUM* IN WOUND HEALING

Isha Gupta[1,2], Sonia Gandhi[1,*], PhD and Sameer Sapra[2], PhD

[1]Metabolomics Research Facility,
Division of Behavioural Neuroscience,
Institute of Nuclear Medicine and Allied Sciences,
Defence Research and Development Organization, Delhi, India
[2]Department of Chemistry, Indian Institute of Technology, Delhi, India

ABSTRACT

Wounds are a global health concern for the world's major population, and it is influenced by many factors such as the local environment of wounds, untreated acute wounds, ageing, diabetes, and arterial disease, which may delay the healing process of wounds. Around 1% of the population suffers from wounds due to the high economy, degradation of

[*]Corresponding Author's E-mail: sonia@inmas.drdo.in.

living standard, low literacy standard, lack of exposure to health facilities, and complex and long treatments. Different wound healing therapies are being investigated and developed by the researchers to reduce the healing time of wounds and the complexity of wounds. Ancient traditional wound healing therapies derived from herbs are also gaining attention due to their cost-effectiveness and non-toxicity. In this chapter, we will discuss *Origanum*, the Turkish spice, which has a history dated back to centuries. This herb has found wide applications due to the presence of primary and secondary metabolites. *Origanum* is used in different forms such as oil, ointment or extract to reduce the complexities in the wound healing process. The essential oil obtained from different species of *Origanum* contains thymol, carvacrol, p-cymene, thymoquinone, ɣ-terpinene and other compounds which are responsible for the healing properties. Study supporting the use of oregano ointment over its essential oil also has been discussed briefly. Researchers have also evaluated the administration of extracts on animal models for its efficacy in wound healing. Due to the goal of sustainable development, traditional therapies will flourish in the coming future.

Keywords: wound, wound healing, traditional therapies, herbs, *Origanum*

INTRODUCTION

Rupturing of the skin can be a threat to the human population. As, the body system is covered by the integumentary system, which consists of nails, hair, skin and glands. In humans, the skin is the largest organ system, responsible for survival, sensation, healing, protection and maintaining thermal homeostasis. The skin has two layers: Epidermis and Dermis. Epidermis covers the whole body surface, which is composed of keratinocytes. Keratinocytes protect the external environment and avoid the loss of water.

In contrast, dermis covers the connective tissue. It provides nourishment and energy to the upper layer. It has elastin, collagen and ground substance. Then, beneath this layer is loose connective tissue, hypodermis. It contains hair roots and flat cells (Basu, Narendra Kumar, and Manjubala 2017; Yousef and Sharma 2017).

The wounds formed due to damage of layers of the skin by any chemical or physical injury (Lordani et al. 2018). The wound is a *silent epidemic*, a global health concern which is affected by ageing and any chronic disease (Gonzalez et al. 2016). These chronic diseases responsible for a wound are venous ulcers, trauma vasculitis, diabetes, tuberculosis, leprosy and atherosclerosis. The cases of type 2 diabetes, metabolic syndrome and peripheral vascular disease are increasing rapidly. All over the world, 300 million and 100 million people suffer from chronic and acute wounds, respectively (Das and Baker 2016). However, in India, 4.5 and 10.5 per 1000 people suffer from chronic and acute wounds respectively. Another cause for the development of chronic wound is the insufficient and improper treatment to the acute wound (Shukla, Ansari, and Gupta 2005).

As the number of the elderly population is increasing, diabetes patients are also increasing. Hence, the wound care sector needs to be more developed to deal with burns, infected, and chronic wound (Shukla, Ansari, and Gupta 2005; Gonzalez et al. 2016). The number of people suffering from any wound is increasing due to the high economy, degradation of living standard, limited knowledge, high cost of health-care, complex and long treatments (Gonzalez et al. 2016). According to Sushruta Samhita, the wound is a "*discontinuity of the skin or other body surface which, when healed, leaves a scar*." If chronic wounds are left untreated may also lead to the comorbidities.In Ayurveda, around 164 medicinal plants have been described for their wound healing applications.

Hippocrates said that "*For an obstinate ulcer, sweet wine and much patience should be enough*." When all over the world, the physicians were discussing wounds, the Greeks never remain backstage. They were the first to distinguish between acute and chronic wounds as 'fresh' and 'non-healing' respectively. Boiled water, vinegar and wine were used for sterilising the wounds. French biologist Louis Pasteur and British surgeon Joseph Lister made breakthroughs by developing the prevention methods by identifying the cause of infection. The founders of Johnson & Johnson began mass production of wound dressing and other surgical products which revolutionised the medical field. Wounds can be open, closed, with or without tissue loss and burn (Paul and Sharma 2015).

This chapter discusses the different wound categories based on involved skin layers and tissue, complexity such as acute, chronic and burn. Then, the three major stages of wound healing: Inflammation, proliferation and migration are discussed in brief. The development in herbal wound healing therapies for involving oregano, neem, tulsi, and curcumin is mentioned to give a new insight into the wound care sector. The new findings related to the use of oregano as ointment, essential oil and extract for the assessment of wound healing in different wound models has been discussed in detail.

TYPE OF WOUND

The wounds are classified into different categories based on the skin layers, tissue involvement, the complexity of wound and tissue loss, illustrated in Figure 1.

The wounds can be open and closed. Abrasion, hematoma and contusion are closed wounds. As, these wounds include damage to sensory nerves, vessels, tissue layers while the skin layers remain unaffected. Later on, pigmentation and atrophy are noticed which delay the process of healing. These wounds are quite painful. Many wounds such as a gunshot, insect bites and their stings, lacerations, metabolic and neurological wounds come under the category of open wounds. Except for lacerations, deep tissues are damaged in the open wounds. These take significant time in healing.

Wounds based on tissue loss are divided into categories—wounds with tissue loss associated with loss of tissue layers. Here, granulation tissue fills the wound area in the healing process. Wounds without tissue loss are formed due to the crushing followed by bleeding from the wound area.

Based on the skin layer, wounds are of three types. Superficial wounds include damage to epidermis layers. In contrast, full-thickness wound involves damage to epidermis, dermis and subcutaneous tissue. Partial-thickness wounds include damage to epidermis as well as dermis (Sezer and Cevher 2011; Basu, Narendra Kumar, and Manjubala 2017).

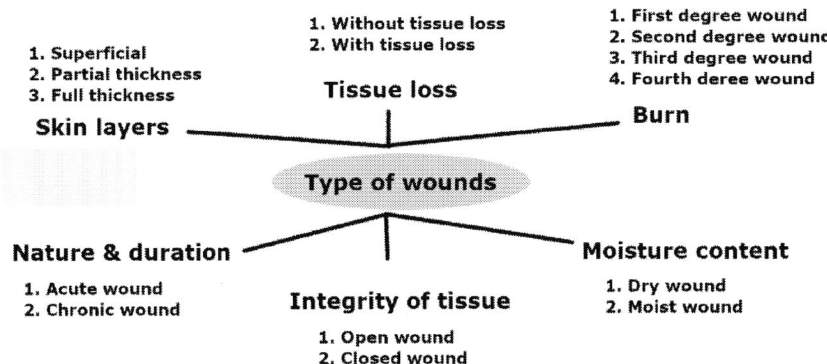

Figure 1. Type of wounds.

Table 1. Different type of acute and chronic wounds

Acute Wound	Chronic Wound
Cuts	Pressure Ulcer
Crush wounds	Arterial and venous ulcer
Missile wounds	Diabetic ulcer
Abrasions	
Contusions	
Avulsion	
Punctures	
Laceration	

Burn: In this type of wound, skin or the organ is damaged by an external agent such as current, flammable chemical or heat. Burns can severely affect the healing process by altering the following events: aggregation of platelets, fibrinolysis, vascular permeability. Burns can be divided based on the involvement of the skin layers. First degrees are caused by the intense sunlight, heat or flame. Oedema in the first-degree burns heals within 24 hours while the wound fully heals within a week. Second-degree burns are deeper than first degree because the epidermal and some of the dermal layers are involved. The second-degree burn may convert to third in the presence of infection, which is caused by short term contact with heat. Second-degree burn can be superficial or deep dermal. In superficial second degree burn, plasma is secreted from the affected area, and recovery takes around four weeks. In deep dermal burns, epidermis and dermis are damaged by the chemicals or electric current. Severe pain and hyperanesthesia are observed. After healing, recovery takes two months with the appearance of the scar. It may sometime lead to the development of the third-degree burn on the persistent exposure to hot water, current and flame. Grafting or granulation followed by contraction is important for decreasing the recovery time. A fourth-degree burn is due to the carbonisation of burned tissue (Sezer and Cevher 2011).

The wound can be divided based on nature and duration of the healing process to acute and chronic wounds, explained in Table 1. Acute wounds are simple injuries that heal in a short span of 8-12 weeks without alteration in the wound healing process sequence. In contrast, chronic wounds are formed on constant tissue contact and heal with alteration in normal phases (Dhivya, Padma, and Santhini 2015; Basu, Narendra Kumar, and Manjubala 2017). The chronic wounds cases are of ulcers in diabetic and heart patients (Lordani et al. 2018). The causes leading to chronic wounds are ageing, diabetes, malnutrition, and poor health.

WOUND HEALING

Wound healing is complex, multiple stages, physiological and psychological process (Sezer and Cevher 2011; Gonzalez et al. 2016). The healing process is well-coordinated and overlapping of regenerative and degenerative processes governed by various cells and their interactions. It is a synergism of different processes of dermal, epidermal cells, extracellular matrix (ECM), growth factors (GF), plasma-derived proteins and an assemblage of cytokines (Ganapathy et al. 2012; Basu, Narendra Kumar, and Manjubala 2017). In prolonged medical conditions, disruption of signalling pathway leads to a chronic wound (Paul and Sharma 2015). Wound healing comprises of three stages which are sequential and overlapping: Inflammation, Proliferation and Remodelling, shown in Figure 2.

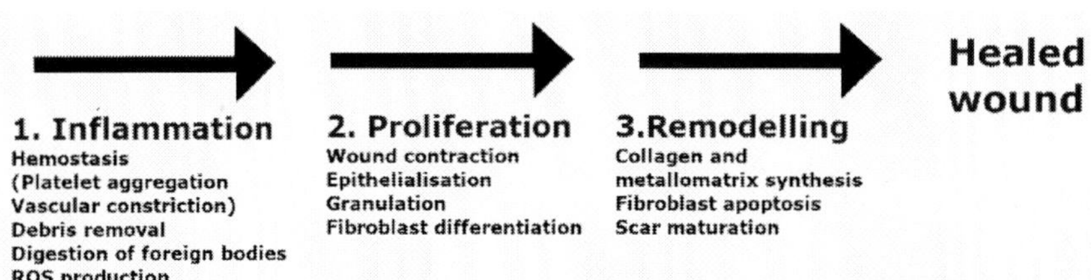

Figure 2. Wound healing.

Inflammation

This stage commences instantly after the loss of tissue to prevent fluid loss and last for six days.Further, it involves hemostasis which is governed by platelets. Hemostasis is the immediate and critical response to excessive bleeding. Here, the wound is sealed, followed by the formation of fibrin by the clotting factors. Fibrin mesh is produced on the site of injury. The platelets induce vasoconstriction. The role of neutrophils is to perform phagocytosis digesting foreign bodies and bacteria. Simultaneously monocytes enter the site and changes to macrophages to release protease and transforming growth factor beta-1 (TGF-β1) which are responsible for migration of epithelial cells and fibroblasts. The platelets and blood cells are trapped on the wound, further strengthening collagen deposition (Maver et al. 2015; Mihai et al. 2019; Shedoeva et al. 2019).

Proliferation

It is a vital step which begins from the 3rd day of the wound to two weeks (Lordani et al. 2018).Moreover, it involves wound contraction, epithelialisation and granulation. In granulation, extracellular matrix (ECM), containing collagen, is also produced in large amount by fibroblast. Then, the keratinocytes differentiate and propagate to form epidermis, which leads to a decrease in the size of the wound. The non-essential cells undergo apoptosis (Maver et al. 2015; Mihai et al. 2019; Shedoeva et al. 2019).

Remodelling

The last stage of wound healing can take considerable time in restoring the structure and functioning of the tissue. It involves the synthesis of collagen and metallomatrix proteins in the dermal layer (Gantwerker and Hom 2012; Maver et al. 2015; Nethi et al. 2019).

The sequence of the healing process is altered by external and internal factors such as comorbidities, necrotic tissue, hypothermia, poor nutrition, age and pathogens. Improper wound healing may lead to spread of infection to the interior organs accelerating the severity of the wound. Therefore, patients frequently suffer from pain, excessive exudates, reduced movement and morbidity (Dhivya, Padma, and Santhini 2015; Shedoeva et al. 2019). Often, biofilm formation at the site of the wound is observed due to the *Escherichia coli*, *Staphylococcus aureus*, and *Pseudomonas aeruginosa* (Mihai et al. 2019). When the balance of these phases is disturbed, the outcome varies. If the duration of inflammation increases, chronic non-healing wounds are developed which are common in diabetic and vascular diseases patients. Suppose the duration of proliferation increases, quality of life is reduced.

Hence, different therapies should be developed by the joint efforts of researchers and clinicians to reduce the unpleasant result such as scarring and to decrease the healing time. The herbal formulations have many advantages compared to synthetic drugs such as cheaper, fewer side effects and shorter wound healing time.

TRADITIONAL WOUND HEALING THERAPIES

Since ancient times, traditional therapies are being used for the treatment of any type of wounds such as acute, chronic and burn. The traditional therapies are obtained from natural herbs such as *Curcuma longa* (turmeric), *Origanum* (Oregano), *Aloe barbadensis* (Aloe vera), *Arcticum lappa* (woodland burdock), *Ampelopsis japonica*, *Rhodiola imbricata*, *Cinnamomum cassia* (cinnamon), *Centella asiatica* (gotu kola), *Celosia argentea* (silver cocksomb), *Camellia sinensis* (tea tree), *Calendula officinalis* (pot marigold), *Caesalpinia sappan* (patamag), *Boswellia sacra* (frankincense), and *Andrographis paniculata* (kariyat). Around 80% of people from regions such as the Middle East, Africa, Asia, and Latin America are dependent on traditional therapies for wound care management because of their affordability and minimal side effects (Maver et al. 2015; Shedoeva et al. 2019). The research related to their wound healing activity has strengthened their application as a therapeutic agent. Therefore, optimisation, standardisation and validation of the natural products need to be done by the health care sector. Approximately 25% of medicines are composed of plant-derived substances (Svoboda and Svoboda 2003). Few therapies from herbs are discussed below concisely.

Aloe barbadensis

This herb is generally known as Aloe vera, belongs to the family of Liliaceae and grown in hot desert regions. It is an endemic species of Africa. Here, Aloe vera is also known as the desert lily. This herb is being used for the treatment of wounds, burns and infections by people of Egypt, Africa, Spanish and Greeks. It has many chemical compounds such as anthracene hydroxyl derivatives (aloins), chromone, sugars (glucose, cellulose), minerals (sodium, magnesium, calcium, copper, zinc), vitamins, and enzymes (catalase, amylase) (Hashemi, Madani, and Abediankenari 2015; Hekmatpou et al. 2019). Kayode used twelve rabbits, created wound on the back of rabbits and then, applied Aloe vera gel. The results proved that the healing rate increased significantly, and changes were observable in the count of lymphocyte and neutrophile, and the volume of the packed cell and mean corpuscular (Kayode 2016).

Avena sativa

Avena sativa is commonly known as oats, avena, ma-karasu-mugi and hafer. It is a cereal species, cultivated in Marmara and Anatolia. As per Lebanon tradition, its ethanolic extract is used to rheumatoid and neuralgia. In Ladakh, it is used for treating kidney and urinary disorders. This plant comprises a significant amount of vitamins, minerals, proteins, lipids, and fibres. Many kinds of research have also proved to benefit many conditions such as inflammation, itchiness, and HIV (Akkol et al. 2011). Akkol et al. examined the wound healing property of *A. sativa* while preparing in different solvents such as water, ethanol, ethyl acetate, and n-hexane. Ethanol extract showed noticeable wound healing property. This type of study supported the traditional use of ethanolic extract for the treatment of skin and inflammation (Akkol et al. 2011). Hosary et al. prepared biopolymer composite from β-glucan obtained from the *A. sativa.* β-glucan, a polysaccharide, known for triggering tissue granulation, biosynthesis and deposition of dermal fibroblast and reepithelialisation (Hosary et al. 2020).

Azadirachta indica

Azadirachta indica is commonly known as neem in India, member of Meliaceae family. It is mostly found in Bangladesh, Nepal, Pakistan and India. The main component found in neem are limonoids, β-sitosterol, quercetin, nimbin, nimbidin and nimbolide. It has many properties, such as anti-inflammatory, anti-microbial, anti-tumour, anti-gastric, anti-arthritic, and anti-pyretic (Alzohairy 2016). According to a study, extract of *A. indica* boost the wound healing process in the incision and excision model of Sprague Dawley rats (Barua et al. 2010). Nagesh et al. assessed the rate of wound contraction in incision and excision model for topical administration of ethanolic extract of *A. indica* (Nagesh, Basavanna, and Kishore 2015). Maan et al. prepared *A. indica* extract in ethanol, water and water-ethanol (1:1 v/v) and then prepared an ointment vehicle made up of macrogol 600 and macrogol 400. These were topically administrated on the dorsal surface of the excision wound model of Male Swiss Albino mice. Only water extract administered mice showed increased wound contraction rate of 93.39% (Maan, Yadav, and Yadav 2017).Singh et al. enrolled 60 patients for the treatment. Three groups were made: neem oil, haridra powder, and both drugs. Neem oil was topically applied and found effective because 43.8% of patients in group I showed 50% wound healing (Singh et al. 2014). Chundran et al. topically applied the ethanolic extract of neem on the excision wound model of *Rattus norvegicus*. The wound healing rate of leaf extract of neem was the same as that of povidone-iodine (Chundran, Husen, and Rubianti 2015).

Centella Asiatic

Centella Asiatic is commonly known as Indian pennywort, gotu kola, jalbrahmi or mandukparni, found in many regions of the Middle East and Asia.This perennial herb grows in moist and damp places. It is used in medicine as a blood purifier, maintaining blood pressure, sedative, anti-depressant, anti-epileptic, anti-oxidant, etc. The main constituents of the jalbrahmi are saponins, brahmoside, brahminoside, centelloside, asiatic acid, and tannins (Gohil, Patel, and Gajjar 2010; Mukherjee et al. 2011). Somboonwong et al. compared the wound healing activities of *Centella asiatic* in different extracts such as water, methanol, ethyl acetate and hexane in partial-thickness burn and incision model. The extract treated groups showed epithelisation and keratinisation in the burn injury model compared to the control and untreated group of Sprague Dawley rats (Somboonwong et al. 2012).

Cinnamon

Cinnamon, used as a spice is *Cinnamomum zeylancium* and *Cinnamomum aromaticum*. Cinnamon shows various properties such as anti-inflammatory, anti-microbial, anti-oxidant, anti-diabetic, anti-tumour. The chemical substances present in cinnamon are eugenol, cinnamaldehyde, cinnamic acid, eugenol, weitherhin, tannins, mucilage, diterpenes, and proanthocyanidins. Farahpour et al. analysed the wound healing effect in the excision rat model on topically applying *Cinnamomum zeylancium*. The thirty-two rats were distributed into four groups, and these rats were kept under observation for 21 days. *C. zeylancium* increased wound healing by increasing the epithelialisation process (Farahpour, Amniattalab, and Hajizadeh 2012). Ahmadi et al. conducted the study to investigate the effect of essential oil of *C. verum*. Three groups of mice were made, and then the full-thickness wounds were developed and induced the biofilm formation by *P. aeruginosa* and *S. aureus*. This herb's topical application increased the proliferation, distribution of fibroblast, epithelialisation, and the synthesis of collagen and keratin. It also reduced the duration of the inflammation phase (Seyed Ahmadi, Farahpour, and Hamishehkar 2019).

Curcuma longa

Curcuma longa L. which is also known as turmeric belongs to the family of Zingiberaceae, found in the regions of Asia and Middle Eastern countries. The curcumin, obtained from this herb, has been used extensively as a home remedy in Indian households for any kind of inflammation and skin disease. It is also used as colouring and flavouring agent in cuisines of India, Indonesia and Chinese. The anti-cancer and anti-ageing properties of curcumin have also been reported (Thangapazham, Sharad, and Maheshwari 2016). This herb has majorly three curcuminoids: curcumin, demethoxycurcumin, and bisdemethoxycurcumin. Curcumin, a polyphenol, yellow coloured pigment, obtained from the rhizome part of this plant. It is known for its anti-oxidant, anti-inflammatory, hypolipidemic, and anti-cancer properties (Tayyem et al. 2006). Curcumin accelerates the tissue remodelling, collagen deposition, granulation and wound closure process. The wound healing of curcumin has been studied in rat and guinea pig wound model. The curcumin at a 40 mg/Kg dose increases the wound process in 7 to 11 days and decreased the healing time by 35% in rats (Thangapazham, Sharad, and Maheshwari 2016).

Mimosa tenuiflora

Mimosa tenuiflora is commonly called as tepescohuite, belongs to the family of Mimosaceae, found in South and Central America. The bark of this shrub is used to treat burns and wound since ancient times. In 1980, the bark got the attention of the scientist. The shrub shows many properties such as anti-microbial, anti-inflammatory, wound healing and cicatrising (Zippel, Deters, and Hensel 2009). Rivera-Arce et al. did a clinical trial on venous leg ulcer patients and administered the hydrogel containing 5% of the extract. After the 8th week of treatment, the ulcer decreased by 92%, and no side-effects were noticed in the patients (Rivera-Arce et al. 2007).

Ocimum sanctum

This aromatic shrub belongs to the basil family of Lamiaceae, originate in India and now grown in many regions of the Eastern tropics. As per the Ayurveda, it is known as 'Mother Medicine of Nature,' 'The Queen of Herbs,' and 'The Incomparable One.' This herb is well known for many applications such as asthma, cough, fever, arthritis, indigestion, eye diseases, diarrhoea, vomiting, cardiac disorders and skin diseases (Cohen 2014). In India, it has medicinal as well as spiritual importance. Shetty et al. evaluated the different solvent extract of *O. sanctum* in rats. They treated rats with different doses of aqueous and ethanolic extract. Both the extracts showed good results such as enhanced hydroxyproline, superoxide dismutase, hexosamines, hexuronic acid, and catalase (Shetty, Udupa, and Udupa 2008).

Origanum

Origanum (Oregano) belongs to Lamiaceae genus have perennial, shrubby and aromatic herbs of around 42 species. These species are majorly found in Mediterranean, Iran-Siberian and Euro-Siberian regions while around 23 species are found in Turkey. These species show anti-oxidant, antimutagenic, anti-inflammatory and anti-microbial properties. Therefore, this genus has been part of traditional Turkish medicines. *Origanum vulgare* L. is the highest quality species of this genus. This species is further divided into six subspecies: *vulgare*, *hirtum*, *viridulum, virens, gracile* and *glandulosum* (Vokou, Kokkini, and Bessiere 1993; S. Kokkini, Karousou, and Hanlidou 2003; Chishti, Kaloo, and Sultan 2013). This species is used as a spice in Eastern and Mediterranean region. Many kinds of research have shown that it possesses antibacterial, anti-genotoxic and anti-fungal properties (Ragi and Amy 2011; Sarikurkcu et al. 2015). The major components present in *Origanum* are thymol and carvacrol (Svoboda and Svoboda 2003). Its oil has a rich amount of sesquiterpenes and monoterpenes which show anti-fungal, anti-oxidant and anti-microbial properties. The protocatechuic compound, found in oregano, inhibits lipid peroxidation and reactive oxygen species (Lordani et al. 2018).

RECENT ADVANCES OF *ORIGANUM* IN WOUND HEALING

Herbs are plants whose different parts are used in perfumery, food, beverage, cosmetic, and pharmaceutical industries. There are 31 plant families, and these have been divided into the six groups based on their principal flavour components. The Labiatae or Lamiaceae family include around 3500 aromatic species, mostly found in South America, South-west Asia and Mediterranean regions. It has culinary herbs which are part of traditional herbal medicine. Some of the examples of aromatic herbs are discussed above. Genus *Origanum* is a versatile medicinal plant of the basil family (Stella Kokkini, Karousou, and Vokou 1994; S. Kokkini, Karousou, and Hanlidou 2003). The Letswaart classified this genus into ten divisions in 1980. It has been part of Turkish medicine. The divisions are *Anatolicon* Bentham, *Amaracus* (Gleditsch) Bentham, *Brevifilamentum* letswaart, *Chilocalyx* (Briquet) letswaart, *Longitubus* letswaart, *Camapanulaticalyx* letswart, *Majorana* (Miller) Bentham, *Elongatispica* letswaart, *Prolaticorolla* letswaart, and *Origanum*. *Origanum vulgare* L. has the widest distribution in this genus. It has many properties such as anti-inflammatory, anti-oxidant, insecticidal, anti-microbial, anti-fungal, anti-spasmodic, anti-tumoral, anthelminthic, anti-parasitic, stimulant, carminative, tonic, diaphoretic, and analgesic. This plant contains more than a hundred compounds comprising origanosides, flavonoids and depsides (Stella Kokkini, Karousou, and Vokou 1994; Chishti, Kaloo, and Sultan 2013).The major component found in the *Origanum* is thymol and carvacrol. The other components are γ-terpinene, ρ-cymene and thymoquinone (Chávez-González, Rodríguez-Herrera, and Aguilar 2016).

Nowadays, researchers are trying to find the evidence justifying the efficacy of *Origanum* in wound healing by performing *in vitro* and *in vivo* studies.

Extract

Plant extracts are obtained by adding any solvent such as water, and ethanol to the part of the plant and then heating it for a certain period followed by filtration. The solvent dependent phytochemicals are extracted in that particular solvent.

Moslemi et al. investigated the wound healing activity of *O. vulgare*. Their team prepared the extract via cold maceration process to evaluate the surgical wound affected by *Staphylococcus aureus*, one of the major Gram-positive bacteria. Twenty-two Sprague Dawley rats were used for this study. The excision wound on the dorsal side was created and inoculated with the bacterial strain. The rats were divided into two groups, where group 1 received the topical administration of *O. vulgare* and group 2 did not receive any such treatment. In the biochemical analysis, hydroxyproline content was higher in the treated rats, which increased collagen content.Hexosamine content was also found higher in treated rats which triggered the granulation process early. Therefore, accelerated wound healing was observed in the topical administration (Moslemi et al. 2015).

Yogesh et al. conducted a study for the evaluation of the wound healing activity of the 5% and 10% ethanolic extract from the flowers of *O. majorana*. These extracts were topically administrated to the incision and excision rat models. In the excision rat model, 92.24% and 84% of wound contracted on 15th day for 10% and 5% respectively compared to 97.73% from nitrofurazone (Yogesh et al. 2016).

Essential Oils

Essential oils are complex, low molecular weight secondary metabolites and volatile liquid synthesised by the parts of plants in response to the attack by any insect, herbivores or other living organisms (Lv et al. 2011; Mihai et al. 2019). These are stored in the ducts of oil and resin, glands and trichomes. In industries, the oils are extracted from the glands with the help of steam, solvent extract or hydrophilisation. The secondary metabolites are known for their regenerative properties because it reduces the oxidative stress, biofilm formation and the inflammation (Mihai et al. 2019). Oils have complex compounds such as terpenoids and phenylpropanoids. Many essential oils such as citrus, rosemary, oregano, basil and mentha are widely available in the market. Oregano oil obtained from the *Origanum* herb showed high efficacy against microorganisms. (Chávez-González, Rodríguez-Herrera, and Aguilar 2016).

Gutierrez et al. emphasised that the higher concentration of the oils is required to achieve the anti-microbial effects. The team used oregano combinations with other essential oils such as basil, rosemary, sage, thyme, lemon balm, and basil to evaluate the anti-microbial effects. In the checkerboard method, all the combinations were effective against the *B. cereus*. As per the findings, the combination of oregano with thyme and majoram was effective against *E. coli* and *P. aeruginosa*. In contrast, oregano with basil has greater activity against both Gram-negative organisms (Gutierrez, Barry-Ryan, and Bourke 2008).

Lv et al. planned a study to evaluate the anti-microbial properties of ten plants' essential oil, including *Origanum vulagre* against four common bacteria and yeasts, and determined the essential oils' chemical composition. Oregano essential oil showed the highest efficacy and broad-spectrum against all microorganisms. The *E. coli* was the most resistant strain which showed moderate inhibition for basil and oregano essential oils. The anti-microbial activity of oregano is due to the presence of carvacrol, thymol, p-cymene, and γ-terpinen. It also showed the lowest minimum inhibition concentration (MIC) at 0.625 μL/mL for all the strains (Lv et al. 2011).

Gunal et al. studied the effect of carvacrol which is derived from *Origanum onites* on excisional skin injury model of male Wistar-Albino rats. The research team applied 12.5% of carvacrol topically on the injury for five days consecutively. On the eighth day, the thickness of granulation tissue was less in carvacrol treated group than the control group. The levels of TNF-α, $IL\text{-}1_{\beta}$ and $TGF\text{-}\beta_1$ were analysed by using ELISA to provide support to the above data. The administration of carvacrol increased the production of pro-inflammatory molecules at the site of injury (Gunal, Heper, and Zaloglu 2014).

Sarikurkcu et al. isolated the essential oils from two subspecies of *Origanum vulgare* (*O. vulgare* subsp. *vulgare* and *O. vulgare* subsp. *hirtum*). They reported that thymol and linalool were the main components in *O. vulgare* subsp. *vulgare* and *O. vulgare* subsp.*hirtum* respectively. The MIC was found to be 85.3 μg/ml for former subspecies against *Sarcinalutea* and latter against *Candida albicans*. The former also showed high reducing power and free radical scavenging property (Sarikurkcu et al. 2015).

Ointment

Topical ointment for wound healing can be used to terminate the bacterial growth on the damaged skin tissue. The bacitracin and neomycin are the most common and synthetic topical ointment. The major drawback of using ointments is the allergic reaction of the skin.

J. et al. recruited forty adult volunteers of age 18-75 years old and randomly divided into two groups: one receiving petroleum jelly ointment and the other receiving oregano ointment. The ointments were applied to the wound twice a day. Only the culture test of 19 per cent showed a positive test for *Staphlococcus aureus* in the oregano treated group compared to 41 per cent of the petroleum-treated group. The one patient of the oregano treated group showed cellulitis as compared to three of the petroleum-treated groups. The other noticeable changes were pigmentation, scar colour and pliability for oregano treated group (J. et al. 2011).

CONCLUSION

Traditional medicine system can give a new shape to the health care sector due to its low cost-effectiveness, simplified preparation protocols, natural repairment of skin and no toxicity. The aromatic herbs belonging to the Lamiaceae family are found all around the world.This family is worldwide known for biological as well as pharmacological applications. *Origanum* is part of this family, being part of Turkish medicines since ancient times. As discussed in the above topics, this herb has many characteristics such as anti-inflammatory, anti-oxidant, insecticidal, anti-microbial, anti-fungal, anti-spasmodic, anti-tumoral, anthelminthic, anti-parasitic, stimulant, carminative, tonic, diaphoretic, and analgesic. Here, we have discussed the application of this herb in different forms such as oil and extract. This herb increases the rate of wound healing in different models by a different mode of actions such as anti-oxidant, anti-microbial, anti-inflammatory, and accelerating granulation process. Therefore, researchers need to investigate, standardise and validate its application to be used commercially on a large scale. They can also isolate, identify and standardise the secondary metabolites such as carvacrol and thymol present in the extract. Then, they can establish data related to their efficacy in this particular field. In this way, the traditional medicine system will provide the proper benefit to the world's population.

REFERENCES

Akkol, E. Küpeli, I. Süntar, I. Erdogan Orhan, H. Keles, A. Kan, and G. Çoksari. 2011. "Assessment of Dermal Wound Healing and in Vitro Antioxidant Properties of *Avena Sativa* L."*Journal of Cereal Science* 53 (3): 285–90. https://doi.org/10.1016/j.jcs.2011.01.009.

Alzohairy, Mohammad A. 2016. "Therapeutics Role of Azadirachta Indica (Neem) and Their Active Constituents in Diseases Prevention and Treatment."*Evidence-Based Complementary and Alternative Medicine* 2016: 1–11. https://doi.org/10.1155/2016/7382506.

Barua, C. C., A. Talukdar, A. G. Barua, A. Chakraborty, R. K. Sarma, and R. S. Bora. 2010. "Evaluation of the Wound Healing Activity of Methanolic Extract of Azadirachta Indica (Neem) and Tinospora Cordifolia (Guduchi) in Rats."*Pharmacologyonline* 1: 70–77.

Basu, Poulami, U. Narendra Kumar, and I. Manjubala. 2017. "Wound Healing Materials - A Perspective for Skin Tissue Engineering." *Current Science* 112 (12): 2392. https://doi.org/10.18520/cs/v112/i12/2392-2404.

Chávez-González, M. L., R. Rodríguez-Herrera, and C. N. Aguilar. 2016. "Essential Oils: A Natural Alternative to Combat Antibiotics Resistance." In *Antibiotic Resistance*, 227–37. Elsevier. https://doi.org/10.1016/B978-0-12-803642-6.00011-3.

Chishti, Shayista, Zahoor A. Kaloo, and Phalestine Sultan. 2013. "Medicinal Importance of Genus *Origanum*: A Review." *Journal of Pharmacognosy and Phytotherapy* 5 (10): 170–77. https://doi.org/10.5897/JPP2013.0285.

Chundran, Naveen Kumar, Ike Rostikawati Husen, and Irra Rubianti. 2015. "Effect of Neem Leaves Extract (Azadirachta Indica) on Wound Healing." *Althea Medical Journal* 2 (2): 199–203. https://doi.org/10.15850/amj.v2n2.535.

Cohen, MarcMaurice. 2014. "Tulsi - Ocimum Sanctum: A Herb for All Reasons." *Journal of Ayurveda and Integrative Medicine* 5 (4): 251. https://doi.org/10.4103/0975-9476.146554.

Das, Subhamoy, and Aaron B. Baker. 2016. "Biomaterials and Nanotherapeutics for Enhancing Skin Wound Healing." *Frontiers in Bioengineering and Biotechnology* 4 (82): 1–20. https://doi.org/10.3389/fbioe.2016.00082.

Dhivya, Selvaraj, Viswanadha Vijaya Padma, and Elango Santhini. 2015. "Wound Dressings – a Review." *BioMedicine* 5 (4): 22. https://doi.org/10.7603/s40681-015-0022-9.

Farahpour, Mohammad Reza, Amir Amniattalab, and Hadi Hajizadeh. 2012. "Evaluation of the Wound Healing Activity of Cinnamomum Zeylanicum Extract on Experimentally Induced Wounds in Rats." *African Journal of Biotechnology* 11 (84): 15068–71. https://doi.org/10.5897/AJB11.1421.

Ganapathy, Nalliappan, SivaSubramaniyan Venkataraman, Rajkumar Daniel, RamrajJayabalan Aravind, and VilapakkamBhikshewaran Kumarakrishnan. 2012. "Molecular Biology of Wound Healing."*Journal of Pharmacy and Bioallied Sciences* 4 (6): 334. https://doi.org/10.4103/0975-7406.100294.

Gantwerker, Eric A., and David B. Hom. 2012. "Skin: Histology and Physiology of Wound Healing." *Clinics in Plastic Surgery* 39 (1): 85–97. https://doi.org/10.1016/j.cps.2011.09.005.

Gohil, Kashmira J., Jagruti A. Patel, and Anuradha K. Gajjar. 2010. "Pharmacological Review on Centella Asiatica: A Potential Herbal Cure-All." *Indian Journal of Pharmaceutical Sciences* 72 (5): 546–56. https://doi.org/10.4103/0250-474X.78519.

Gonzalez, Ana Cristina De Oliveira, Tila Fortuna Costa, Zilton De Araújo Andrade, and Alena Ribeiro Alves Peixoto Medrado. 2016. "Wound

Healing - A Literature Review." *Anais Brasileiros de Dermatologia* 91 (5): 614–20. https://doi.org/10.1590/abd1806-4841.20164741.

Gunal, Mehmet Y., Aylin O. Heper, and Nezahat Zaloglu. 2014. "The Effects of Topical Carvacrol Application on Wound Healing Process in Male Rats." *Pharmacognosy Journal* 6 (3): 10–14. https://doi.org/10.5530/pj.2014.3.2.

Gutierrez, J., C. Barry-Ryan, and P. Bourke. 2008. "The Antimicrobial Efficacy of Plant Essential Oil Combinations and Interactions with Food Ingredients." *International Journal of Food Microbiology* 124 (1): 91–97. https://doi.org/10.1016/j.ijfoodmicro.2008.02.028.

Hashemi, Seyyed Abbas, Seyyed Abdollah Madani, and Saied Abediankenari. 2015. "The Review on Properties of Aloe Vera in Healing of Cutaneous Wounds." *BioMed Research International* 2015: 714216. https://doi.org/10.1155/2015/714216.

Hekmatpou, Davood, Fatemeh Mehrabi, Kobra Rahzani, and Atefeh Aminiyan. 2019. "The Effect of Aloe Vera Clinical Trials on Prevention and Healing of Skin Wound: A Systematic Review." *Iranian Journal of Medical Sciences* 44 (1): 1–9. http://www.ncbi.nlm.nih.gov/pubmed/30666070.

Hosary, Rania EL, Shereen M.S. EL Mancy, Kadriya S. EL Deeb, Hanaa H. Eid, Mona E. EL Tantawy, Manal M. Shams, Rasha Samir, Nouran H. Assar, and Amany A. Sleem. 2020. "Efficient Wound Healing Composite Hydrogel Using Egyptian Avena Sativa L. Polysaccharide Containing β-Glucan." *International Journal of Biological Macromolecules* 149 (xxxx): 1331–38. https://doi.org/10.1016/j.ijbiomac.2019.11.046.

Kayode, Olaifa Abayomi. 2016. "Effects of Aloe Vera Gel Application on Epidermal Wound Healing in the Domestic Rabbit." *International Journal of Research in Medical Sciences* 5 (1): 101. https://doi.org/10.18203/2320-6012.ijrms20164531.

Kokkini, S., R. Karousou, and E. Hanlidou. 2003. "HERBS | Herbs of the Labiatae." *Encyclopedia of Food Sciences and Nutrition*, 3082–90. https://doi.org/10.1016/b0-12-227055-x/00593-9.

Kokkini, Stella, Regina Karousou, and Despina Vokou. 1994. "Pattern of Geographic Variations of *Origanum Vulgare* Trichomes and Essential Oil Content in Greece." *Biochemical Systematics and Ecology* 22 (5): 517–28. https://doi.org/10.1016/0305-1978(94)90046-9.

Lordani, Tarcisio Vitor Augusto, Celia Eliane de Lara, Fabiana Borges Padilha Ferreira, Mariana de Souza Terron Monich, Claudinei Mesquita da Silva, Claudia Regina Felicetti Lordani, Fernanda Giacomini Bueno, Jorge Juarez Vieira Teixeira, and Maria Valdrinez Campana Lonardoni. 2018. "Therapeutic Effects of Medicinal Plants on Cutaneous Wound Healing in Humans: A Systematic Review." *Mediators of Inflammation* 2018: 1–12. https://doi.org/10.1155/2018/7354250.

Lv, Fei, Hao Liang, Qipeng Yuan, and Chunfang Li. 2011. "In Vitro Antimicrobial Effects and Mechanism of Action of Selected Plant Essential Oil Combinations against Four Food-Related Microorganisms." *Food Research International* 44 (9): 3057–64. https://doi.org/10.1016/j.foodres.2011.07.030.

Maan, Preeti, Kuldeep Singh Yadav, and Narayan Prasad Yadav. 2017. "Wound Healing Activity of Azadirachta Indica A. Juss Stem Bark in Mice." *Pharmacognosy Magazine* 13 (50): S316–20. https://doi.org/10.4103/0973-1296.210163.

Maver, Tina, Uroš Maver, Karin Stana Kleinschek, Dragica M. Smrke, and Samo Kreft. 2015. "A Review of Herbal Medicines in Wound Healing." *International Journal of Dermatology* 54 (7): 740–51. https://doi.org/10.1111/ijd.12766.

Mihai, Mara Madalina, Monica Beatrice Dima, Bogdan Dima, and Alina Maria Holban. 2019. "Nanomaterials for Wound Healing and Infection Control." *Materials* 12 (13): 2176. https://doi.org/10.3390/ma12132176.

Moslemi, Hamid-reza, Meysam Tehrani-sharif, Saeed Mohammmadpour, Alireza Makhmalbaf, Khatereh Kafshdouzan, Tannaz Ahadi, and Ramin Mazaheri Nezhad Fard. 2015. "Wound Healing Activity of *Origanum Vulgare* against Surgical Wounds Infected by

Staphylococcus Aureus in a Rat Model." *Iranian Journal of Veterinary Medicine* 9 (2): 135–42. https://doi.org/10.22059/ijvm.2015.54012.

Mukherjee, Sourav, Swapnil Dugad, Rahul Bhandare, Nayana Pawar, Suresh Jagtap, Pankaj K. Pawar, and Omkar Kulkarni. 2011. "Evaluation of Comparative Free-Radical Quenching Potential of Brahmi (Bacopa Monnieri) and Mandookparni (Centella Asiatica)." *AYU (An International Quarterly Journal of Research in Ayurveda)* 32 (2): 258–64. https://doi.org/10.4103/0974-8520.92549.

Nagesh, H. N., P. L. Basavanna, and M.S. Kishore. 2015. "Evaluation of Wound Healing Activity of Ethanolic Extract of Azadirachta Indica Leaves on Incision and Excision Wound Models in Wister Albino Rats." *International Journal of Basic and Clinical Pharmacology* 4 (6): 1178–82. https://doi.org/10.18203/2319-2003.ijbcp20151354.

Nethi, Susheel Kumar, Sourav Das, Chitta Ranjan Patra, and Sudip Mukherjee. 2019. "Recent Advances in Inorganic Nanomaterials for Wound-Healing Applications." *Biomaterials Science* 7 (7): 2652–74. https://doi.org/10.1039/c9bm00423h.

Paul, Willi, and Chandra P. Sharma. 2015. *Advances in Wound Healing Materials: Science and Skin Engineering. Smithers Rapra*. https://doi.org/2162-1934.

Ragi, J., Pappert A., Rao B., Havkin-Frenkel D., and Milgraum S. 2011. "Oregano Extract Ointment for Wound Healing: A Randomised, Double-Blind, Petrolatum-Controlled Study Evaluating Efficacy." *Journal of Drugs in Dermatology* 10 (10): 1168–72. http://jddonline.com/articles/download_article/1623%5Cnhttp://ovidsp.ovid.com/ovidweb.cgi?T=JS&PAGE=reference&D=emed10&NEWS=N&AN=2011578222.

Ragi, Jennifer, and Pappert Amy. 2011. "Oregano Extract Ointment for Wound Healing:" *Journal of Drugs in Dermatology* 10 (10): 1168–72.

Rivera-Arce, Erika, Marco Antonio Chávez-Soto, Armando Herrera-Arellano, Silvia Arzate, Juan Agüero, Iris Angélica Feria-Romero, Angélica Cruz-Guzmán, and Xavier Lozoya. 2007. "Therapeutic Effectiveness of a Mimosa Tenuiflora Cortex Extract in Venous Leg

Ulceration Treatment." *Journal of Ethnopharmacology* 109 (3): 523–28. https://doi.org/10.1016/j.jep.2006.08.032.

Sarikurkcu, Cengiz, Gokhan Zengin, Mustafa Oskay, Sengul Uysal, Ramazan Ceylan, and Abdurrahman Aktumsek. 2015. "Composition, Antioxidant, Antimicrobial and Enzyme Inhibition Activities of Two *Origanum Vulgare* Subspecies (Subsp. *Vulgare* and Subsp. Hirtum) Essential Oils." *Industrial Crops and Products* 70: 178–84. https://doi.org/10.1016/j.indcrop.2015.03.030.

Seyed Ahmadi, Seyed Gharani, Mohammad R. Farahpour, and Hamed Hamishehkar. 2019. "Topical Application of Cinnamon Verum Essential Oil Accelerates Infected Wound Healing Process by Increasing Tissue Antioxidant Capacity and Keratin Biosynthesis." *The Kaohsiung Journal of Medical Sciences* 35 (11): 686–94. https://doi.org/10.1002/kjm2.12120.

Sezer, Ali Demir, and Erdal Cevher. 2011. "Biopolymers as Wound Healing Materials: Challenges and New Strategies."*Biomaterials Applications for Nanomedicine*, 383–414.

Shedoeva, Aleksandra, David Leavesley, Zee Upton, and Chen Fan. 2019. "Wound Healing and the Use of Medicinal Plants." *Evidence-Based Complementary and Alternative Medicine* 2019 . https://doi.org/10.1155/2019/2684108.

Shetty, Somashekar, Saraswati Udupa, and Laxminarayana Udupa. 2008. "Evaluation of Antioxidant and Wound Healing Effects of Alcoholic and Aqueous Extract of Ocimum Sanctum Linn in Rats." *Evidence-Based Complementary and Alternative Medicine* 5 (1): 95–101. https://doi.org/10.1093/ecam/nem004.

Shukla, V. K., Mumtaz A. Ansari, and S. K. Gupta. 2005. "Wound Healing Research: A Perspective from India." *The International Journal of Lower Extremity Wounds* 4 (1): 7–8. https://doi.org/10.1177/1534734604273660.

Singh, Anjali, Anil Kumar Singh, G Narayan, Teja B Singh, and Vijay Kumar Shukla. 2014. "Effect of Neem Oil and Haridra on Non-Healing Wounds." *AYU (An International Quarterly Journal of*

Research in Ayurveda) 35 (4): 398. https://doi.org/10.4103/0974-8520.158998.

Somboonwong, Juraiporn, Mattana Kankaisre, Boonyong Tantisira, and Mayuree H. Tantisira. 2012. "Wound Healing Activities of Different Extracts of Centella Asiatica in Incision and Burn Wound Models: An Experimental Animal Study." *BMC Complementary and Alternative Medicine* 12 (103): 1–7. https://doi.org/10.1186/1472-6882-12-103.

Svoboda, K P, and T Svoboda. 2003. *HERBS| Herbs and Their Uses*, no. 1970: 3071–77.

Tayyem, Reema F, Dennis D Heath, Wael K Al-Delaimy, and Cheryl L Rock. 2006. "Curcumin Content of Turmeric and Curry Powders." *Nutrition and Cancer* 55 (2): 126–31.

Thangapazham, Rajesh L., Shashwat Sharad, and Radha K. Maheshwari. 2016. "Phytochemicals in Wound Healing." *Advances in Wound Care* 5 (5): 230–41. https://doi.org/10.1089/wound.2013.0505.

Vokou, Despina, Stella Kokkini, and Jean Marie Bessiere. 1993. "Geographic Variation of Greek Oregano (*Origanum Vulgare* Ssp. Hirtum) Essential Oils." *Biochemical Systematics and Ecology* 21 (2): 287–95. https://doi.org/10.1016/0305-1978(93)90047-U.

Yogesh, Anjana, Latha M, Saravana Balaji MD, Premlal KR, Vinod Mony, and Sengottuvelu S. 2016. "Wound healing activity of monoterpene rich *Origanum majorana* on experimentally induced wound in rats." *International Journal of Phytopharmacology* 7 (4): 233–36.

Yousef, Hani, and Sandeep Sharma. 2017. *Anatomy, Skin (Integument), Epidermis.*

Zippel, Janina, Alexandra Deters, and Andreas Hensel. 2009. "Arabinogalactans from Mimosa Tenuiflora (Willd.) Poiret Bark as Active Principles for Wound-Healing Properties: Specific Enhancement of Dermal Fibroblast Activity and Minor Influence on HaCaT Keratinocytes." *Journal of Ethnopharmacology* 124 (3): 391–96. https://doi.org/10.1016/j.jep.2009.05.034.

INDEX

A

B

C

M

N

O

P

Q

R